Applied Geochemistry with Case Studies on Geological Formations, Exploration Techniques and Environmental Issues

Edited by Luis Felipe Mazadiego, Eduardo De Miguel Garcia, Fernando Barrio-Parra and Miguel Izquierdo-Díaz

Published in London, United Kingdom

IntechOpen

Supporting open minds since 2005

Applied Geochemistry with Case Studies on Geological Formations, Exploration Techniques and Environmental Issues
http://dx.doi.org/10.5772/intechopen.74885
Edited by Luis Felipe Mazadiego, Eduardo De Miguel Garcia, Fernando Barrio-Parra and Miguel Izquierdo-Díaz

Contributors
Gaafar El Bahariya, Wanilson Luiz-Silva, Alice Bosco-Santos, Jiang Chen, Trolard Fabienne, David Kaniewski, Guilhem Bourrié, Tomás Martín-Crespo, David Gómez-Ortiz, Silvia Martín-Velázquez, Shaun Williams, Wei-Yu Shi

First published in London, United Kingdom, 2020 by IntechOpen
IntechOpen is the global imprint of INTECHOPEN LIMITED, registered in England and Wales, registration number: 11086078, 7th floor, 10 Lower Thames Street, London, EC3R 6AF, United Kingdom
Printed in Croatia

British Library Cataloguing-in-Publication Data
A catalogue record for this book is available from the British Library

Additional hard and PDF copies can be obtained from orders@intechopen.com

Applied Geochemistry with Case Studies on Geological Formations, Exploration Techniques and Environmental Issues
Edited by Luis Felipe Mazadiego, Eduardo De Miguel Garcia, Fernando Barrio-Parra and Miguel Izquierdo-Díaz
p. cm.
Print ISBN 978-1-78985-884-6
Online ISBN 978-1-78985-915-7
eBook (PDF) ISBN 978-1-78985-916-4

For EU product safety concerns:
IN TECH d.o.o., Prolaz Marije Krucifikse Kozulić 3, 51000 Rijeka, Croatia, info@intechopen.com or visit our website at intechopen.com.

Meet the editors

Luis Felipe Mazadiego is a mining engineer, doctor, and professor at the Universidad Politécnica de Madrid (UPM). He teaches several master's degree subjects (mining engineering, energy engineering, research, modeling and risk analysis in the environment, oil and gas engineering, teacher training) and is oriented, in many cases, the planning and management of projects. For his research on the applications of surface geochemistry, he obtained the Extraordinary Doctorate Award and the Juan Artieda Award. He is an elected member of INHIGEO, a scientific association dependent on UNESCO. He has participated in research projects on environment, mining heritage, shale gas, and natural radioactivity and published more than 100 scientific articles and congress communications.

Since he joined Universidad Politécnica de Madrid in 1990, Eduardo De Miguel Garcia has pursued three main research lines: geochemical prospecting, environmental geochemistry, and human health risk analysis. His research career began with the development of his PhD thesis in collaboration with the German Geological Survey (Bundesanstalt für Geowissenschaften und Rohstoffe, BGR) and the Norwegian Atmospheric Research Institute (Norsk Institutt for Luftforskning, NILU). He remains in close collaboration with prestigious international researchers in the field of urban geochemistry and human health risk assessment. He has led 9 research projects with a total funding of nearly 500000 € and has had fruitful collaborations with regional environmental authorities and private companies. His teaching activity (in the fields of geochemistry, soil pollution and environmental risk assessment) is not limited to Universidad Politécnica de Madrid, in which he is full professor, but he has also taught outside of Spain at Saint Louis University, Jönköping High School, and Linköping University.

Fernando Barrio-Parra is an assistant professor at the School of Mining Engineering of the Polytechnic University of Madrid. He has a bachelor's degree in environmental sciences, a master's degree in research, modelling and environmental risk analysis, and a PhD in natural resource conservation. His research is related to the modelling of environmental processes, covering the human-environment interaction, littoral geodynamics, geochemical and soil and groundwater pollution processes and prospecting, including human health and environmental risk assessment. His teaching is related to chemistry, geochemistry, and environmental risk assessment.

Miguel Izquierdo-Díaz is a teaching assistant at the School of Mining and Energy Engineering of the Technical University of Madrid. He has a BSc in environmental sciences, a MSc in research, modelling and environmental risk analysis, and his PhD was focused on the study of atmospheric pollution phytomonitoring, probabilistic risk assessment, and trace element bioaccessibility in urban gardens. He is a member of the research group PROMEDIAM (Prospecting and Environment Laboratory). His research is focused on soil and groundwater pollution modelling, biomonitoring and remediation; environmental and human health risk assessment; urban geochemistry; and sustainable agriculture. He also takes part in several European educational projects to bring science and critical raw materials closer to high school students.

Contents

Preface **XI**

Chapter 1 **1**
Geochemistry and Tectonic Setting of Neoproterozoic Rocks
from the Arabian-Nubian Shield: Emphasis on the Eastern Desert of Egypt
by Gaafar A. El Bahariya

Chapter 2 **27**
Tsunami Elemental Signatures in the Samoan Islands: A Case Study
by Shaun Williams

Chapter 3 **39**
Soil Carbon Biogeochemistry in Arid and Semiarid Forests
by Wei-Yu Shi, Xiao-Cong Zhu, Feng-Bao Zhang, Kai-Bo Wang, Lei Deng and Ming-Guo Ma

Chapter 4 **53**
The Geochemical Data Imaging and Application in Geoscience: Taking the Northern Daxinganling Metallogenic Belt as an Example
by Jiang Chen, Zhaoxia Mao, Yan Zhao, Chunpeng Zhang and Shan Jiang

Chapter 5 **71**
Geochemical Methods to Assess Agriculture Sustainability
by Trolard Fabienne, Kaniewski David and Bourrié Guilhem

Chapter 6 **85**
Geoenvironmental Characterization of Sulfide Mine Tailings
by Tomás Martín-Crespo, David Gómez-Ortiz and Silvia Martín-Velázquez

Chapter 7 **111**
Implications of Sediment Geochemistry and Diet Habits in Fish Metal Levels and Human Health Risk
by Alice Bosco-Santos and Wanilson Luiz-Silva

Preface

Geochemistry has always been a multidisciplinary scientific field, characterized by bringing together and synthesizing knowledge and tools from numerous related subjects. It is also a dynamic field of knowledge which, from its formal origins in the 19th century to the present day, has encountered and confronted new challenges. Without abandoning research in mature fields, such as mineral exploration or the geochemical characterization of tectonic processes, the interest of geochemistry has been directed, in recent decades, to some of the most imminent challenges for our society, such as the effects and possible mitigation of climate change or the impact of our society's habits on the health of people and ecosystems. As has happened since the beginning of this scientific discipline, the development of these new fields of geochemistry is linked to advances in the available instrumental techniques for an increasingly precise characterization of the distribution of elements, substances, mineral species, and isotopes in the different geological and environmental matrices. All the aspects mentioned above about geochemistry are reflected in this book, in which researchers from five continents make use of advanced analytical techniques (i.e. Mössbauer spectroscopy, Itrax X-Ray Fluorescence core scanning), and tools and concepts of nearby or auxiliary disciplines (geophysics, remote sensing, biology) to advance the knowledge of complex geochemical problems.

We would like to thank all contributors greatly for their collaboration and effort in producing this book.

Luis Felipe Mazadiego, Eduardo De Miguel Garcia,
Miguel Izquierdo-Díaz and Fernando Barrio-Parra
Universidad Politécnica de Madrid,
Escuela Técnica Superior de Ingenieros de Minas y Energía,
Madrid, Spain

Chapter 1

Geochemistry and Tectonic Setting of Neoproterozoic Rocks from the Arabian-Nubian Shield: Emphasis on the Eastern Desert of Egypt

Gaafar A. El Bahariya

Abstract

The Neoproterozoic rocks of the Eastern Desert (ED) of Egypt represent the northwestern part of the Arabian-Nubian Shield (ANS), which was formed during the Pan-African orogenic cycle (950–450). Geochemistry of the different rock units has clarified their compositional variations, tectonic settings, and origins. The ages of these rock units were reported to predict the crustal evolution of the ANS. Island arc volcanic rocks and ophiolitic sequences formed between 700 and 800 Ma, and then, they were obducted in the earlier stage of the Pan-African orogeny. The post-collision stage was characterized by the emplacement of large masses of Dokhan volcanics (610–560 Ma) and shallow level A-type granites (610–550 Ma). Neoproterozoic ophiolites fall geochemically and tectonically into two separate groups: MORB-like ophiolites and SSZ ophiolites of fore-arc tectonic setting. Intra-oceanic island arcs and related inter-arc volcaniclastic sediments are followed by the incorporations of ophiolite fragments into the volcaniclastic matrix to form "ophiolitic mélange" through tectonic and/or concurrent sedimentary and tectonic processes. The "gneissic domes" that are metamorphic core complexes were previously interpreted to represent a pre-Neoproterozoic. However, recent age data argued that the ED gneissic rocks are juvenile in origin and Neoproterozoic. Granitoid rocks in the ED include older and younger types. Most of the older granitoids are of I-type character, displaying metaluminous, calcalkaline geochemical characteristics plot in the area of volcanic arc granites (VAG), whereas most of the younger granitoids are mainly alkaline of A-type granites and of within-plate tectonic setting (WPG). Nonmetamorphosed Dokhan volcanics and Hammamat molasse sediments formed during the final post-collisional phases.

Keywords: geochemistry, Neoproterozoic, Arabian-Nubian Shield, Eastern Desert, Egypt, age dating, crustal evolution

1. Introduction

The Arabian-Nubian Shield (ANS) forms one of the largest exposures of juvenile continental crust (1000–525 Ma) on Earth [1]. It consists of mainly juvenile

IntechOpen

Neoproterozoic crust, now widely exposed in parts of Egypt, Saudi Arabia, Sudan, Eritrea, Ethiopia, Yemen, and Somalia. The ANS was formed during the Neoproterozoic between 900 and 550 Ma through the accretion of intra-oceanic arcs during the closure of the Mozambique Ocean and the amalgamation of Gondwana [2]. These accretion processes led to the formation of well-defined arc-arc and continent-arc suture zones [3, 4]. The ANS was essentially stable continental crust by Early Cambrian time at 530 Ma [5]. The ANS and its surroundings has been the object of geologic investigations for a wide range of geological economic and scientific reasons.

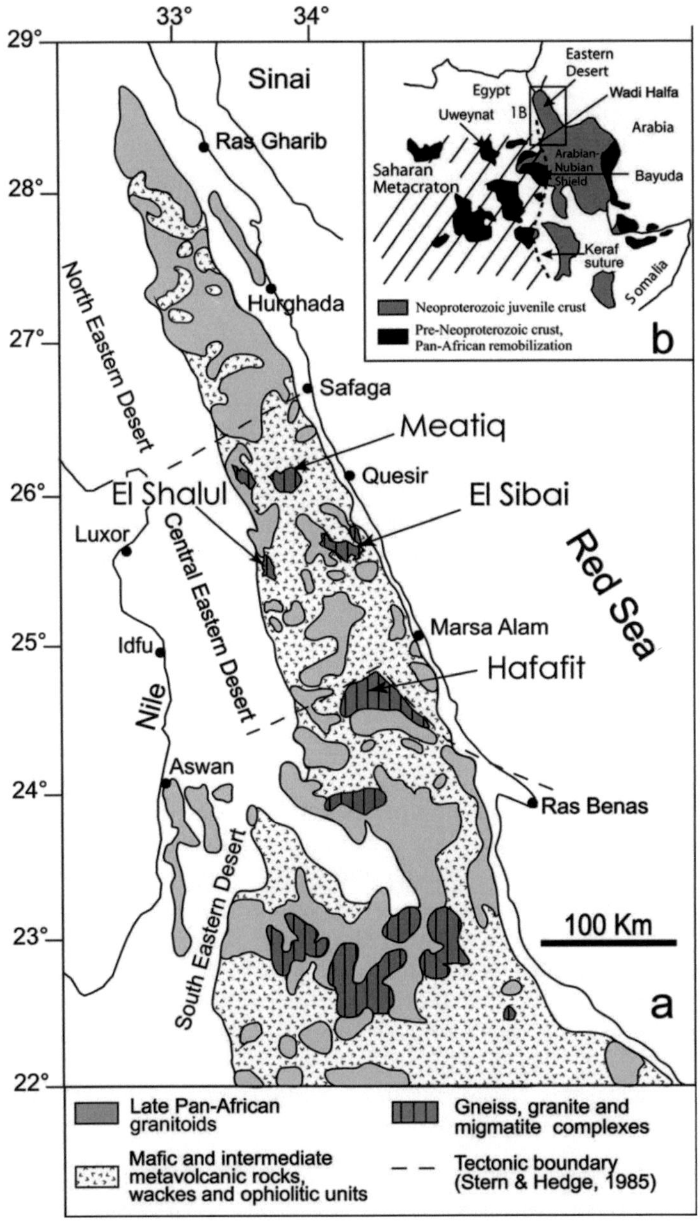

Figure 1.
(a) Inset geological sketch map of NE Africa showing the Arabian-Nubian Shield, the Saharan Metacraton, and Archaean and Palaeoproterozoic crust that was remobilized during the Neoproterozoic and (b) geological map of the Eastern Desert of Egypt showing study areas [15].

The Precambrian basement of the Eastern Desert of Egypt (ED) is a part of the Arabian-Nubian Shield (ANS) and are exposed mainly in the Eastern Desert and Sinai (**Figure 1**). The Eastern Desert of Egypt comprises variably deformed and metamorphosed volcanic, plutonic, and sedimentary rocks of Precambrian age, unconformably overlain by Cretaceous sediments. The Eastern Desert of Egypt is classified into three domains: north, central, and south [6], all revealing different aspects of the region's protracted and intense Neoproterozoic episode of deformation and igneous activity. The Central Eastern Desert (CED) preserves the oldest (Tonian-Cryogenian) history and also best preserves Ediacaran deformation as well as associated (Hammamat) basins. The Southern Eastern Desert (SED) lacks BIF, Ediacaran sedimentary or volcanic successions such as the Hammamat Group and Dokhan Volcanics, whereas the CED does not [7]. The Northern Eastern Desert (NED) is very different than either the CED or the SED. Dokhan volcanics and Hammamat molasses sediments are of widespread occurrence, whereas ophiolites are absent and gneisses are rare.

The reconstructions of this chapter are based on a compiled data of geochemistry and obtained ages on the rock units constituting the Eastern Desert of Egypt. Geochemical data are based on combination of major elements, trace elements, rare earth element (REE) distributions as well as isotope data. In this contribution, I build on previous geological and geochemical studies on the different rock units forming the ED for many years of research to summarize the most important geochemical characteristics of the different rock units and to provide some important information regarding the geochemical dynamic and evolution of Neoproterozoic crust of the ED. This chapter reviews the scope of current geochemical and isotopic datasets for the ANS, with particular emphasis on the Eastern Desert of Egypt.

2. Geologic overview of the Eastern Desert of Egypt

The Precambrian basement rocks of Egypt constitute the northern part of the so-called the Arabian Nubian Shield (ANS), which are exposed mainly in the Eastern Desert (ED) and the Sinai Peninsula (**Figure 1**). The general geological settings of the rock assemblages of the CED were grouped into two major lithotectonic units [8]. The structurally lower one, the "infrastructure," is composed of gneisses and migmatites that crop out in dome structures, such as the Meatiq, Sibai, and Hafafit domes. The overlying unit, the "suprastructure," includes the Neoproterozoic ophiolite complexes and island arc-related metavolcanic and metasedimentary rocks. The suprastructure is also known as the Pan-African Nappe Complex [9]. The juvenile crust in the ED of Egypt is characterized by four main rock units: (i) a gneiss assemblage that comprises the core complexes, (ii) an ophiolite-island arc assemblage, (iii) granitoid intrusions, and (iv) nonmetamorphosed to weakly metamorphosed Dokhan volcanics and Hammamat molasses sediments that unconformably overlie the suprastructure in places [10]. Most of the rock sequences are generally deformed and metamorphosed due to the Neoproterozoic East African orogeny. Below, the most important geological aspects of the main rock units are briefly summarized:

2.1 Granite gneisses and migmatites

A number of medium- to high-grade core complexes or "gneissic domes" have been described in the ED. These infrastructures consist of upper amphibolite facies gneisses, amphibolites, migmatites as well as granitic gneisses. They exposed in several places in the ED, including the Meatiq, El Shalul, the Migif-Hafafit, and

Beitan domes [11–15]. They are generally surrounded by low-grade supracrustal assemblages, and the contact between superstructure and infrastructure is sometimes an intrusive contact and sometimes a high-strain mylonitic zone [16]. The Meatiq Dome consists of Um Baanib deformed granite (cataclastic gneissose granite) forming the core of the dome, followed outward by schists with variable degrees of intercalated amphibolites, together with local mylonites along thrust faults [13]. The Neoproterozoic migmatitic rock association at Wadi Abu Higlig in the Hafafit region is composed of diatexites and schlieric granites (foliated or gneissic granite) in the core of a domal structure flanked by metatexites and preserved amphibolites and metagabbros [14].

2.2 Ophiolite-island arc assemblages

The CED and SED of Egypt are characterized by the widespread distribution of Neoproterozoic ophiolite, ophiolitic mélanges, and intra-oceanic island arc metavolcanic assemblages, along with volcaniclastic metasediments and banded iron formations (e.g., [17–20]). Locally, nearly complete ophiolitic sequences can be observed including serpentinized peridotites, gabbros, sheeted dykes, pillow lavas, and deep-sea sedimentary rocks such as in Ghadir, Muweilih, Esel, El Sid areas [21]. Sheeted dikes are only locally preserved in some localities, whereas pillowed metabasalts are widespread. The ophiolitic peridotites are almost completely serpentinized and are typically altered to talc-carbonate and quartz-carbonates (listwanite) bodies along shear zones. El Bahariya [20] classified the Neoproterozoic ophiolites of the Central Eastern Desert of Egypt based on field geology and mode of occurrences, together with compiled geochemical data into three types: (i) intact MORB ophiolites, (ii) dismembered ophiolites (dismembered blocks and fragments within the mélanges and ophiolites along structural contacts), and (iii) arc-associated ophiolites. The best preserved and nearly intact MORB ophiolites are represented by Wadi Ghadir and Muweilih ophiolites. The arc-associated ophiolite sequences are exposed in Abu Dahr ophiolite, Esel, and El Sid occurrences. Dismembered ophiolites occur either as individual blocks and sheets tectonically emplaced along tectonic contacts or as blocks and fragments within a sheared matrix of volcaniclastic metasediments or metapyroclastics forming "ophiolitic mélange" [19].

The island arc assemblages are concentrated mainly in the CED and SED. They include:

i. metamorphosed volcanic island-arc assemblage and

ii. metamorphosed bimodal volcanic island arc assemblage.

The metamorphosed volcanic island arc assemblages are widespread in the CED and SED [22–24]. They are composed of metavolcanics and related volcaniclastic metasediments. The metavolcanics include metabasalts, metandesites, metadacites, metarhyodacites, and metarhyolites, together with their metapyroclastic counterparts. The volcaniclastic metasediments comprise meta-mudstones, metasiltstones, metagreywackes, metaconglomerates, and schists. The volcaniclastic metasediments together with the metapyroclastics constitute the matrix of the "ophiolitic mélange" [19]. The exotic fragments within melanges are mainly ophiolites of variable sizes and shapes, which include serpentinite and metamorphosed ultramafic rocks, metagabbros, pillowed and massive metabasalts, and minor sheeted dykes and pelagic sedimentary rocks. El Bahariya [19] documents different occurrences of Neoproterozoic ophiolitic melanges in the CED of Egypt and classified the ophiolitic mélanges into:

(i) tectonic mélange, (ii) olistostrome, and (iii) olistostromal mélange. Ophiolitic melanges are also recorded in different occurrences in the SED such as Atshan Ophiolite, Gerf, and Abu Dahr [25].

The bimodal metamorphosed island arc assemblage comprises mafic and felsic volcanic intercalations and arc-related volcanoclastics sediments. They are regionally metamorphosed up to the greenschist facies, locally transformed into schists and amphibolites and commonly associated with banded iron formations and massive sulfides [13, 26, 27]. The metavolcanics together with intra-arc volcaniclastic metasediments occur in different localities in the CED and northern part of SD such as Um Khariga and metapyroclastics, Sodamine, Um Samuky, and El Shadly metavolcanics. The Shadli metavolcanics host some polymetallic massive sulfide mineralizations, e.g., Um Samiuki and Abu Gurdi [28].

2.3 Granitoid rocks

The granitoid rocks constitute about 50% of the basement complex of Egypt. They can, in general, be classified into older and younger granitoids based on their composition, color, and relative age [29]. The older granitoids (~850–635 Ma) comprise trondhjemites, tonalites, granodiorites, and rarely granites, whereas the younger granitoids (~630–540 Ma) are predominated by granites and alkali feldspar granites [6]. The younger granites are further classified according to their geological setting and petrography [30] into: (i) phase I granodiorites with minor monzogranites, (ii) phase II (monzogranites and syenogranites), and phase III (alkali feldspar granites). Recently, part of the Younger granites (phase III) are classified as A-type granites [31].

2.4 Nonmetamorphosed rocks

2.4.1 The Dokan volcanic rocks

The later stage of the crustal evolution of the NED and CED is characterized by the eruption of the Dokhan volcanics, which typically include basaltic andesite, andesite, dacite, and rhyolite, together with tuffs, ignimbrite, and agglomerates [32].

2.4.2 Hammamat sediments

The best exposures of the Hammamat molasses sediments found in Wadi Hammamat area of the CED of Egypt [29], where the sedimentary rocks unconformably overlie other old rock units and consist of unmetamorphosed thick sequences of unsorted conglomerates, sandstones, and siltstones. Most of the Hammamat fragments were derived from the Dokhan volcanics and their thickness varies between 4000 m in Wadi Hammamat and 7500 m thick in the Kareim basin. Locally, the Hammamat sediments are sheared and metamorphosed [33].

3. Geochemistry

The compiled available chemical data from of the ED of Egypt are used for the purpose of understanding the geochemistry of Neoproterozoic rocks, and to clarify their geochemical characteristics and tectonic settings. The overall geochemical characteristics of the different rock units are presented as follows.

3.1 Granite gneisses and migmatites

The Hafafit granitic gneisses are enriched in REE, whereas the Um Baanib orthogneiss presents alkaline granite Rear Earth Elements (REE) pattern (**Figure 2a** and **b**). Um Baanib deformed granites (granite gneisses) are enriched in High Field Strength Elements (HFSE) (Zr, Nb, Y, Th), Rb, Ga, and total REE and depleted in MgO, CaO, and V, showing alkaline and A-type characters, whereas Hafafit granitic gneisses are calcalkaline and of I-type granites [34, 35]. In terms of the Nb, Y, and Rb contents, the Hafafit granite gneisses plot in the field of volcanic arc granites [36], whereas the Um Baanib granite gneisses plot within the field of anorogenic or within-plate A-type granites. Aswan orthogneisses are clacalkaline I-type granitoids [37] that are generally described as subduction-related granitoids [38].

Thermobarometry based on composition of coexisting mineral pairs for granite gneisses indicates that peak metamorphism and partial melting occurred at ~750°C and ~5 kb at high H_2O activity for the metatexite. The granite gneiss in the core of Hafafit dome is suggested to have been formed by syntectonic partial melting of lower to middle crustal protoliths [14]. Plagioclase, clinopyroxene, hornblende, garnet, and biotite show compositional variability as a consequence of the composition of protoliths and prevailing P-T conditions of metamorphism (**Figure 2c–e**). Migmatitic rocks provide an example of the close relation among metamorphism, deformation, and melt generation and emplacement. This migmatitic rock association is interpreted as syntectonic anatectic migmatites formed during compressional phase in an Andean-type continental margin tectonic setting.

3.2 Ophiolite-island arc assemblages

3.2.1 Ophiolite geochemistry

The HFSE and (REE) of Neoproterozoic ophiolites of ED of Egypt suggest either similarities with normal-type mid-ocean ridge basalts (N-MORB) or back-arc basin

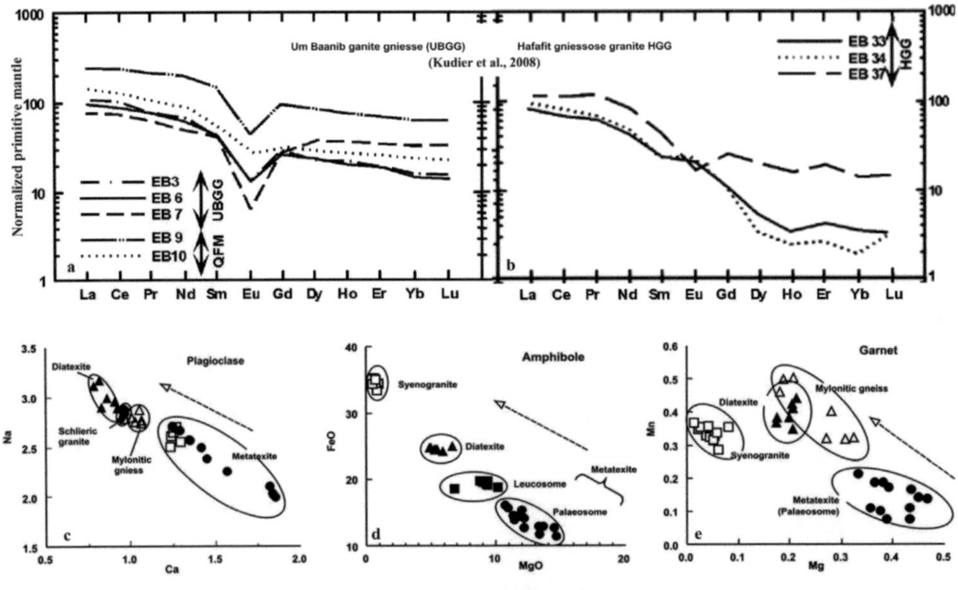

Figure 2.
Rare earth element abundances in the infrastructural rocks from Meatiq (a) and Hafafit core complexes (b) normalized to primitive mantle from [35] and (c–e) compositional variations of plagioclase, amphibole, and garnet in Hafafit migmatitic rocks from [14].

(BAB) magmas or similar to fore arc, boninites and SSZ basalts. Immobile trace-element abundances, together with significant Light Rear Earth Elements (LREE) depletion to almost flat REE patterns for pillow lavas and sheeted dykes of Gerf ophiolite, are compatible with the N-MORB distribution patterns [39]. Volcanic rocks of Fawakhir (El Sid) SSZ ophiolites display moderately depleted to slightly enriched LREE patterns (**Figure 3a**), whereas pillow lavas of Ghadir MORB ophiolites have similar chondrite-normalized REE patterns (**Figure 3b**) [40, 41]. They are enriched in LREE. Most Gerf gabbros have REE patterns with a slight LREE enrichment and a small positive Eu anomaly, whereas the Gerf serpentinized peridotites have Large Ion Lithophile Elements (LILE)-depleted patterns. The Abu Dahr metagabbro and metabasalt have enrichment LILE and LREE enrichment, whereas serpentinized harzburgite and dunite are characterized by enrichment of LILE and nearly flat and unfractionated chondrite-normalized pattern indicating they originated by up to ~30% partial melting of a spinel lherzolite mantle in a subarc setting [25].

Generally, most samples of ophiolitic lavas are subalkaline and reveal tholeiitic affinities, together with minor calcalkaline characters, although subordinate, amount of boninites have been identified as in El Sid ophiolite. On the Ti-V tectonic setting discrimination diagram (**Figure 3c**), generally, ophiolitic metavolcanics and metagabbros of the ED of Egypt fall into two groups: (i) MORB ophiolites and (ii) fore arc or suprasubduction zone (SSZ) ophiolites (e.g., [20, 25, 39, 40–46]). The MORB affinity of metagabbros from Muweilih is documented for the first time by El Bahariya [43], and the whole Muweilih ophiolite sequence is mapped and recorded for the first time as MORB intact ophiolite by El Bahariya [20].

The serpentinites and serpentinized peridotite ophiolites display a diverse suite of geochemical signatures, which make their origin or tectonic setting controversial.

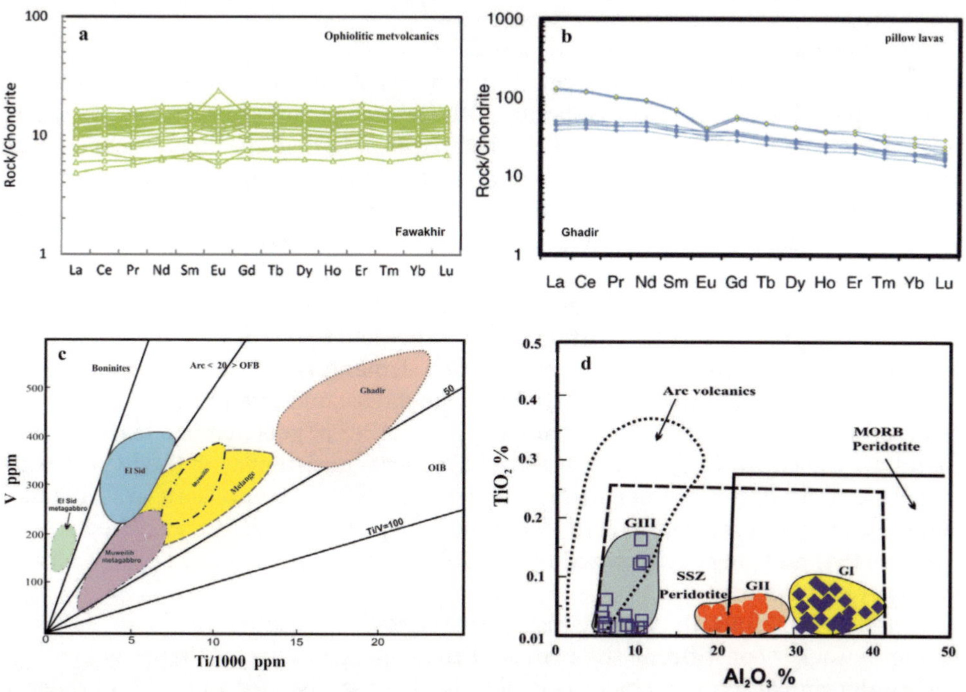

Figure 3.
Geochemical characteristics of ophiolites. (a) Chondrite-normalized REE patterns for Fawakhir (El Sid) ophiolitic pillow lavas, (b) Ghadir pillow lavas [40, 41], (c) Ti/1000 vs. V diagram [20], and (d) chrome spinels from ophiolitic blocks of metamorphosed ultramafics in mélanges [20].

Generally, the chrome spinels from the serpentinites and metamorphosed ultramafic ophiolites have a wide range of Cr#, where the Cr# ranges from ~0.3 to 0.85 and display both MORB and SSZ affinities [45]. They are classified into three groups (G1, G2, and G3) according to their Cr# (**Figure 3d**). Most serpentinized peridotites of the ED show significantly more Mg-rich olivine and chrome spinel with high Cr# (G1 and G2), suggesting a forearc or SSZ environment [42, 45, 47, 48]. Only, data of Cr-spinel from the serpentinized peridotite blocks of Esel olistostrome commonly show low Cr# (G1), and accordingly, they show MORB affinity similar to abyssal peridotites [45]. Moreover, the previous studies dealt collectively with the ophiolitic serpentinites of the ED to be of fore arc or SSZ geochemical signature. However, El Bahariya [20, 47] reported the presence of both SSZ and MORB ophiolitic serpentinized peridotites.

3.2.2 Geochemistry of island arc assemblages

Geochemistry of intermediate and acidic island arc metavolcanics, together with the native intermediate and acidic metavolcanic clasts of the ophiolitic mélanges, is presented. The metavolcanic rocks at Wadi E Dabbah show slightly fractionated REE patterns (**Figure 4a**) and negative Eu and Ce anomalies [49]. The island arc metavolcanics are of oceanic island arc affinity (**Figure 4d**) [23, 24]. The intermediate and acidic island-arc rocks at Gebel Zabara area are calcalkaline and of continental island-arc setting, representing an intermediate maturity stage between the primitive arc and the mature active continental margin [50]. Um Anab meta-andesites, metafelsites, and metarhyolites varieties are predominantly of calcalkaline nature, enriched in LILE and depleted in HFSE, with a pronounced negative Nb anomaly [51]. These rocks are most probably derived from a mantle source produced in an island arc environment where fall in the plate margin field confirming the orogenic nature of these rocks.

The REE patterns of bimodal Um Samiuki metavolcanics rhyodacites are very nearly flat (**Figure 4b**) [28]. Also, the REE patterns of the felsic lavas are slightly LREE-depleted, whereas basalt is slightly LREE-enriched and characterized by negative Eu anomalies. The trace element characteristics of both mafic and felsic members of the Shadli Metavolcanics indicate that these rocks were originated in a magmatic rift. The bimodal metavolcanics at Wadi Sodmien show mafic tholeiitic character and felsic rocks calcalkaline affinity (**Figure 4c**) [52]. They have transitional tectonic setting between island arc/active continental margin and within plate (extensional environment) tectonic setting (**Figure 4d**). Their petrogenesis can be attributed to partial melting of continental crust, and they suggested to be formed in ensialic back arc basin due to extensional rifting. Major trace elements and REE indicate that Igla Eliswid-Um Khariga bimodal mafic and felsic metavolcanic assemblages [53] are clearly tholeiitic in character and share a large number of geochemical features of island-arc tholeiites. The geochemical data are most consistent with the hypothesis that these rocks originated in a magmatic rift. The REE concentrations of Gebel El Hadid banded iron formation (BIF) have LREE depleted and HREE enriched patterns [54] and are characterized by low ΣREE contents (13.7–77.5 ppm) with an average of 45.2 ppm.

On the other hand, the geochemistry of arc-related volcaniclastic metagreywackes constituting the matrix of the mélange indicates that they are chemically similar to quartz-poor oceanic island arc sandstones and were derived mainly from intermediate and felsic volcanic igneous provenances [19, 23, 24]. They are of oceanic island arc tectonic setting (**Figure 4e** and **f**) and appear to be deposited in back-arc basins or interarc basins.

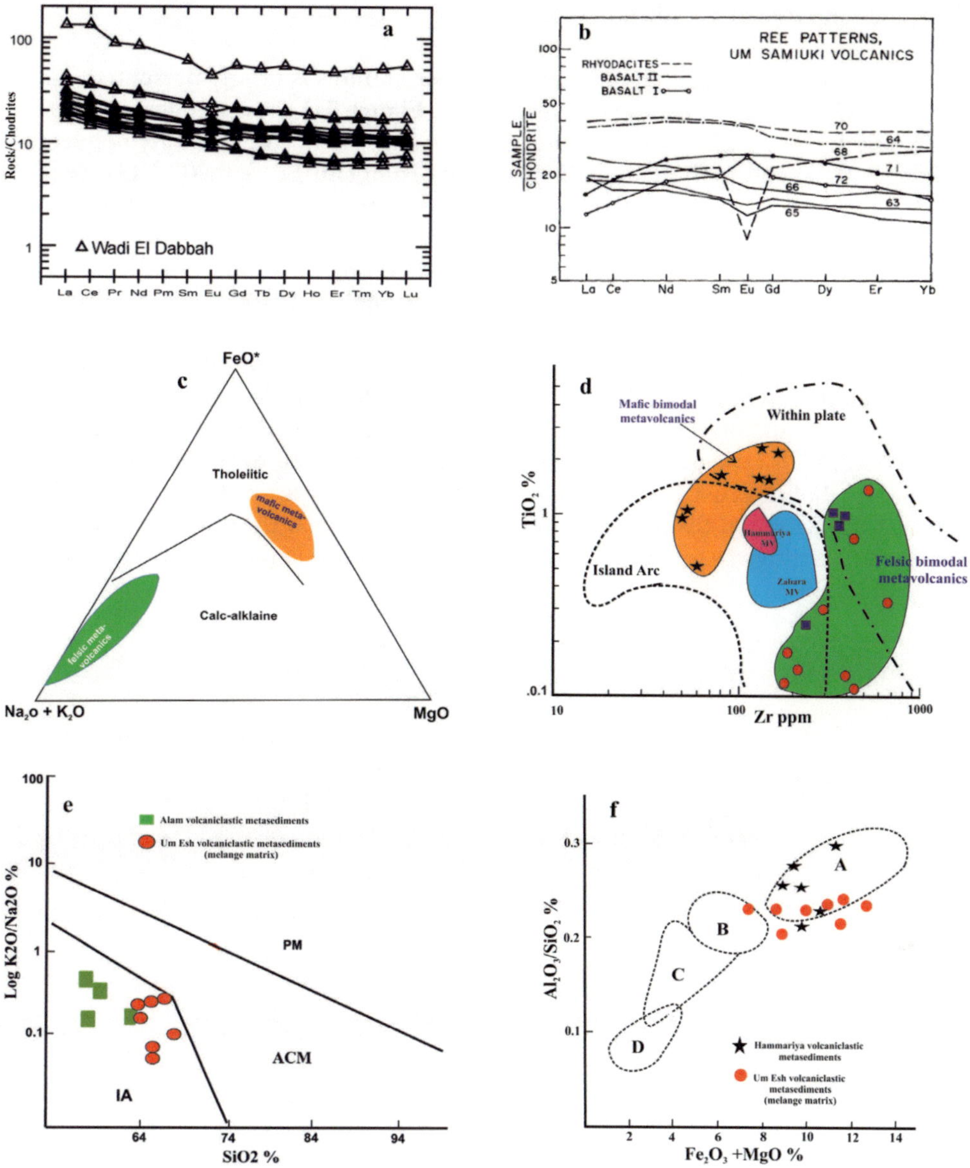

Figure 4.
(a) Rare earth element (REE) and trace element diagrams for the analyzed metavolcanic samples from Wadi El Dabbah from [49]; (b) REE patterns for Um Samiuki Volcanics, normalized to chondritic meteorites from [28]; (c) AFM diagram of Sodmien bimodal metavolcanics, fields based on data from [52]; (d) Sodmien bimodal metavolcanics, data for field of Zabara metavolcanics from [50] and field of Hammariya metavolcanics from [24]; and (e and f) tectonic setting of metagreywackes from matrix of mélanges and from bimodal intra-arc volcaniclastic metasediments, data from [23, 24, 26].

The intra-arc metagreywackes of Alam volcaniclastic metasediments show variable abundances of Zr, Cr, Ni, and V. Their provenance components are mainly of evolved felsic and mafic (bimodal) island arcs and show oceanic arc tectonic setting (**Figure 4e**). They are comparable with Archaean Ranebennur metagreywackes derived from a mixed provenance consisting of mafic and felsic source rocks (e.g., [26, 55]). The rocks are suggested to be deposited in a localized intra-arc basin. The clasts and grains constituting the sediments simulate the principal bimodal volcanic rocks of both the Sukkari metavolcanics and Um Khariga metapyroclastics in the near area [26].

3.3 Geochemistry of granitoid rocks

The geochemistry of both older and younger granites is briefly presented.

The REE patterns of the *older granodiorites* (**Figure 5a**) show enrichment in the LREE relative to HREE, Lan/Ybn values vary from 7.08 to 35.21 (mostly between 7.08 and 19.37) and with Eu anomalies ranging from (Eu/Eu* = 0.70–1.13) [56]. The slightly concave HREE pattern of some biotite suggests hornblende fractionation. *The younger alkali feldspar granites* are characterized by LREE-enrichment (Lan/Ybn = 5.28 – 13.46), moderately fractionated LREE, flat heavy REE patterns (**Figure 5a** and **b**), and moderately to strongly negative Eu anomalies (Eu/Eu* = 0.14–0.63).

Geochemistry of *older granites* reveals that they are metaluminous to slightly peraluminous and have calcalkaline affinity. The older granite can be classified as I-type granites and of volcanic-arc-granite tectonic setting (**Figure 5c**) (e.g., [57]). In the ANS, the I-type granitoids were generally interpreted to result from melting of an amphibolitic crust (e.g., [58]). Moreover, older I-type granites can form through fractionation from mantle-derived, LILE-enriched basaltic melts in subduction settings (e.g., [59]), or from remelting of mafic to intermediate igneous lower crust [60].

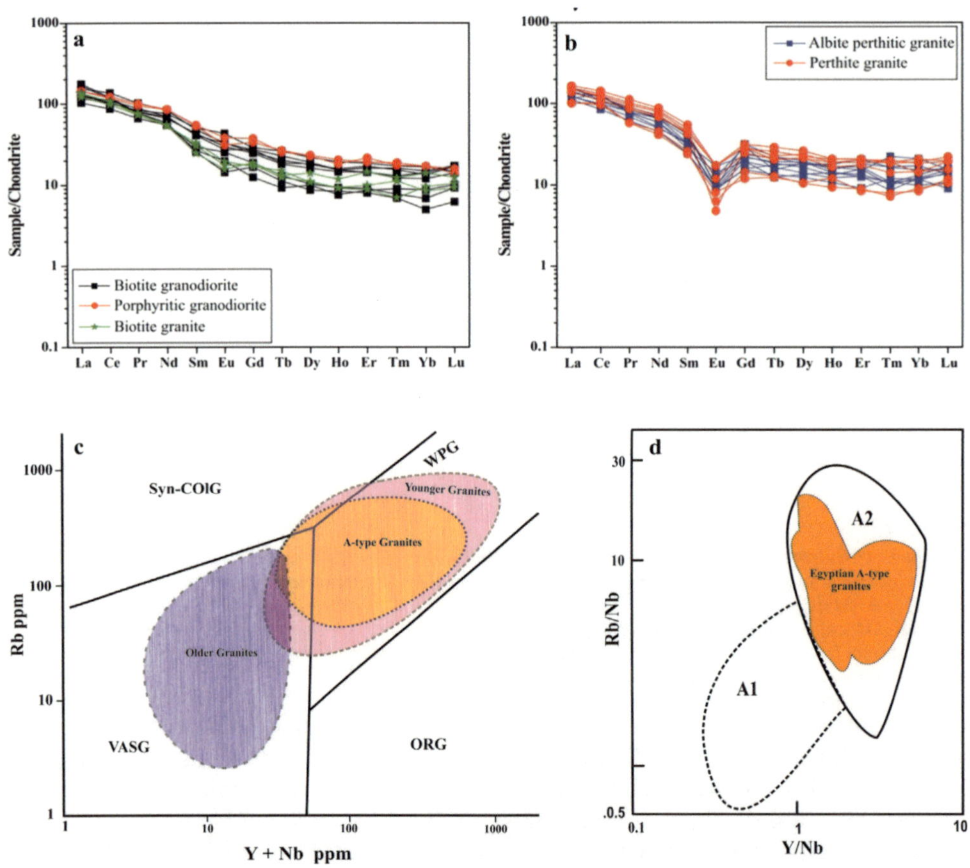

Figure 5.
(a) REE of older granites and (b) REE of younger and A-type granites (from [56]); (c) tectonic setting of granitoid rocks using diagram of Pearce et al. [61], data of older and younger granitoids from [38]; (d) division of A-type granites [62], field of Egyptian A-type granites from combined data from [37, 60] and references therein.

Most of the *younger granites* are LILE-enriched calcalkaline to mildly alkaline rocks commonly of A-type signatures. The younger granites, except phase I exhibit within plate tectonic setting [61], due to their high contents of HFS elements (**Figure 5c**). The phase-III younger granitoids (A-type) are characterized by higher SiO_2, Rb, Y, and Nb and lower MgO, Sr, and Ba contents than other phases of younger granites [38]. The A-type granites are enriched in SiO_2, Na_2O^+ K_2O, FeO*/MgO, Ga/Al, Zr, Nb, Ga, Y, Ce, Rb, and REE and low in CaO, MgO, Ba, and Sr. They are classified as alkaline, and peralkaline to mildly peraluminous A-type granites (e.g., [37]). They are generally enriched in Rb (104–198 ppm), Nb (27–53 ppm), Y (35–79 ppm), Zr (348–750 ppm), and Ga (21–29 ppm), compared to average continental crustal rocks. The overall geochemical characteristics of the A-type granitic rocks of the ED and Sinai are consistent with a within-plate tectonic settings (**Figure 5c**). The A-type granites are eligible for A1-A2 discrimination diagrams after [62] and classified mainly as A2 types (**Figure 5d**), implying that the A-type granites formed mainly in a post-collisional setting. The alkaline A-type granites are generally regarded as the product of either extensive fractional crystallization of mantle-derived mafic magmas (e.g., [63]) or partial melting of various crustal sources (e.g., [37, 49]).

3.4 Geochemistry of nonmetamorphosed rocks

3.4.1 Geochemistry of Dokan volcanic rocks

The Dokan volcanic rocks display well-defined major and trace element trends and a continuum in composition with wide ranges in SiO_2 (54–76%), CaO (8.19–0.14%), MgO (6.96–0.04%), Sr. (983–7 ppm), Zr (328–95 ppm), Cr (297–1 ppm), and Ni (72–1 ppm). The rocks are enriched in LILEs (Rb, Ba, K, Th, Ce) relative to HFSE (Nb, Zr, P, Ti) and have high total REEs with LREE enriched and display variable degrees of enrichment according to rock type (**Figure 6a**) [64]. The intermediate volcanics are characterized by moderate total REE and moderately fractionated patterns with slightly negative Eu-anomalies. Similarly, the REE pattern for the rhyolites is almost identical but with relatively lower content of REE. Generally, the Dokhan volcanics have steep LREE and nearly flat HREE and the large negative Eu anomalies in the rhyolite rocks than those of other varieties indicating formation under condition of relatively low temperature and pressure and/or low water content in the melt.

The geochemistry of the Dokhan volcanic rocks indicates medium-K to high-K calcalkaline affinity, and their tectonic setting is suggested to be: (i) subduction related [65], (ii) extensional setting/rift system (e.g., [66]), and (iii) transitional stage between subduction and extension (e.g., [67, 68]). However, the Dokan lavas mostly plot in an overlap zone between the volcanic arc and within-plate settings on the binary SiO_2—Nb diagram of Pearce and Gale [69] (**Figure 6b**), suggesting a transitional tectonic setting.

3.4.2 Geochemistry of Hammamat molasse sediments

It is of great importance to assess the composition and nature of the source rocks of the Hammamat molasses sediments geochemically, and to determine their tectonic settings. The HFSE are incompatible during most igneous processes; therefore, they tend to be enriched in felsic relative to mafic rocks. Also, they are generally resistant to changes during weathering and alteration processes [70]. The greywackes of the Hammamat molasses sediments have relatively high Zr, Nb, Y, and TH and relatively low Cr, Ni and V, and Sc. **Figure 7a** shows that Um Hassa

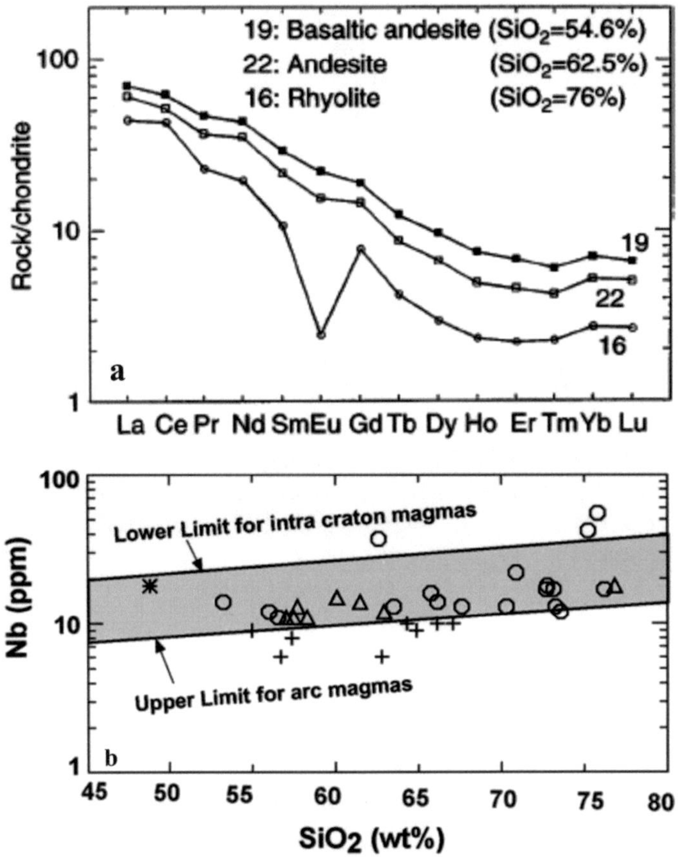

Figure 6.
(a) Chondrite-normalized REE patterns for the Wadi Fatira Dokan volcanics from [64]; (b) SiO_2 vs. Nb diagram after [69] for Wadi Um Sidra and Um Asmer Dokan volcanics from [68].

greywackes of Hammamat molasses sediments have LREE-enriched chondrite-normalized patterns similar to post-Archean Australian shale (PAAS) and UCC patterns [71]. Upper continental crust-normalized patterns for the Um Hassa greywackes reveal significant enrichment of Cr (234–434 ppm) and Ni (49–72 ppm) but depletions in Nb (4.1–7.7 ppm), Rb (33–63 ppm), and Th (3.64–8.92 ppm) relative to UCC values (35, 20, 25, 112, and 10.7 ppm, respectively).

The shale is enriched in REE relative to the coarser sediments (**Figure 7b**), but has a markedly greater Eu anomaly. Chondrite-normalized Ce/Yb ratios are very similar for the shale, the siltstone, and the sandstone ((Ce/Yb)n = 9.8−11.0) [72]. The relatively high K_2O (3.0%), Rb (79 ppm), Ba (1014 ppm), and LREE-enriched pattern ((Ce/Yb)n = 10.3) indicate that the rocks were derived from an LIL and LREE-enriched source. Plausible candidates for this enriched source include the Dokhan volcanics and the Pink younger granite, both of which occur as clasts in the conglomerates and breccias.

There is a close relationship between the tectonic setting of depositional basins and the geochemical characteristics of their sandstones [73–75]. The greywackes from Hammamat molasses sediments plot within the field of active continental margin or continental island arcs (**Figure 7c** and **d**) and appear to be formed in pull-apart intermontane basins of continental margin [33, 71]. The source rocks of the Hammamat molasses sediments are represented mainly by calcalkaline to alkaline felsic source of evolved magmatic island arcs and active continental margin together with minor inputs from calcalkaline island arcs or mafic rocks [33].

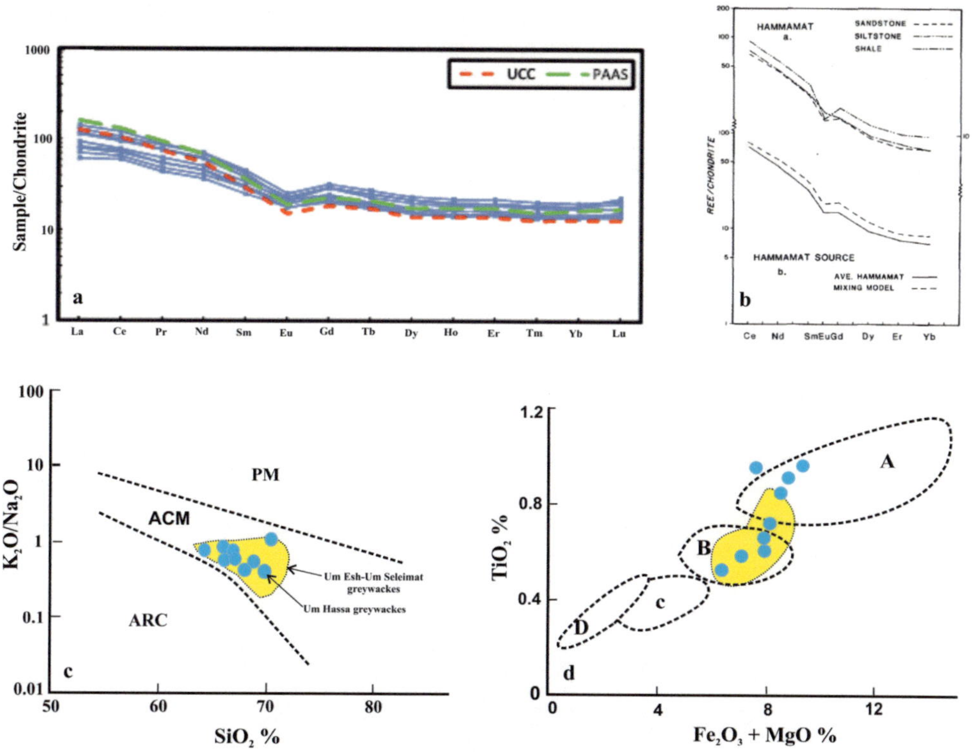

Figure 7.
(a) Chondrite-normalized REE patterns for Um Hassa greywackes from [71]; (b) REE patterns of the Hammamat lithologies from [72]; (c) K_2O/Na_2O vs. SiO_2 after [75] and (d) TiO_2 vs. Fe_2O_3 + MgO after [73] (field of Um Esh-Um Seleimat Hammamat sediments based on data from [33], plots of Um Hassa greywackes from [71].

However, there are minor inputs from island arcs and mafic rocks or ophiolites as reworked clasts from the oldest rocks or from the mélange. The Hammamat molasse area appears to have been deposited in a retroarc foreland basin [71] or appear to be accumulated in intermountain basins or foreland molasse basins [76, 77].

4. Age dating and crustal evolution

Um Ba'anib gneissose granites in the core of Meatiq dome dated 626 [78] and 631 Ma [10]. Ali et al. [15] obtained a zircon age of 631 ± 6 Ma for El-Shalul granitic gneiss. Kröner et al. [36] reported single zircon evaporation ages of 677 ± 9 and 700 ± 12 Ma for granitoid gneisses from the Hafafit gneiss complex and 704 + 8 Ma for migmatitic granitic gneiss from Wadi Bitan. Magmatic emplacement ages for samples from Wadi Beitan yielded 719 ± 10, 725 ± 9 and 744 ± 10 Ma, indicating that the gneiss protoliths are Neoproterozoic [2].

The ophiolitic rocks of the ED have isotopic ages range from 890 to 690 Ma, documenting a 200 Ma year period of oceanic magmatism [79]. The Gerf ophiolites seem to be formed at 741 ± 21 [80], 750 [41], and 730–750 [79]. The ages of the well-preserved ophiolitic rocks in Wadi Ghadir (746 ± 19 Ma, [80]) and in Fawakhir (736.5 ± 1.2 Ma [10]) in the CED are compatible with the ~750 Ma crust forming event proposed by [49].

Stern and Hedge [6] date ED island-arc volcanics to 720–770 Ma. The mafic and felsic lavas of Shadli island arc metavolcanics yield Rb-Sr isochron age of 712 Ma that probably represents the time of volcanic eruption [28]. Ali et al. [49] reported a protolith age of *c*. 750 Ma for the volcanic and volcanosedimentary rocks, and they considered that both the ophiolitic and island arc assemblages in the CED constitute an artifact of one (~750 Ma) crust-forming event.

The emplacement of the Egyptian late- to post-tectonic younger granites covers a time span between 600 and 550 Ma, [6] and 600 and 475 Ma [15]. The underformed Um Had granite has a U-Pb zircon age of 590 ± 3.1 Ma [10]. Some alkaline A-type granites in the NED of Egypt (Al-Missikat, Abu Harba, and Gattar) dated ca. ~600 Ma [81]. Available isochron Rb—Sr ages of alkaline granites are from Sinai fall in the range 550–600 Ma [82].

The transition in the tectonic style from compressional to strong crustal extension is at approximately 600 Ma [66]. Breitkreuz et al. [83] reported age range between 592 and 630 Ma (early Ediacaran) for acidic Dokan volcanics indicating that Dokan volcanism occurred over a 40 Ma time span.

Most of the Eastern Desert molasse basins were evolved between 650 and 580 Ma in individual basins with different individual tectonic settings (e.g., [84]). Rb-Sr whole-rock analyses give an age of 585 ± 15 Ma that approximates the time of sedimentation [75]. U—Pb dating of clastic zircons from the Hammamat group at

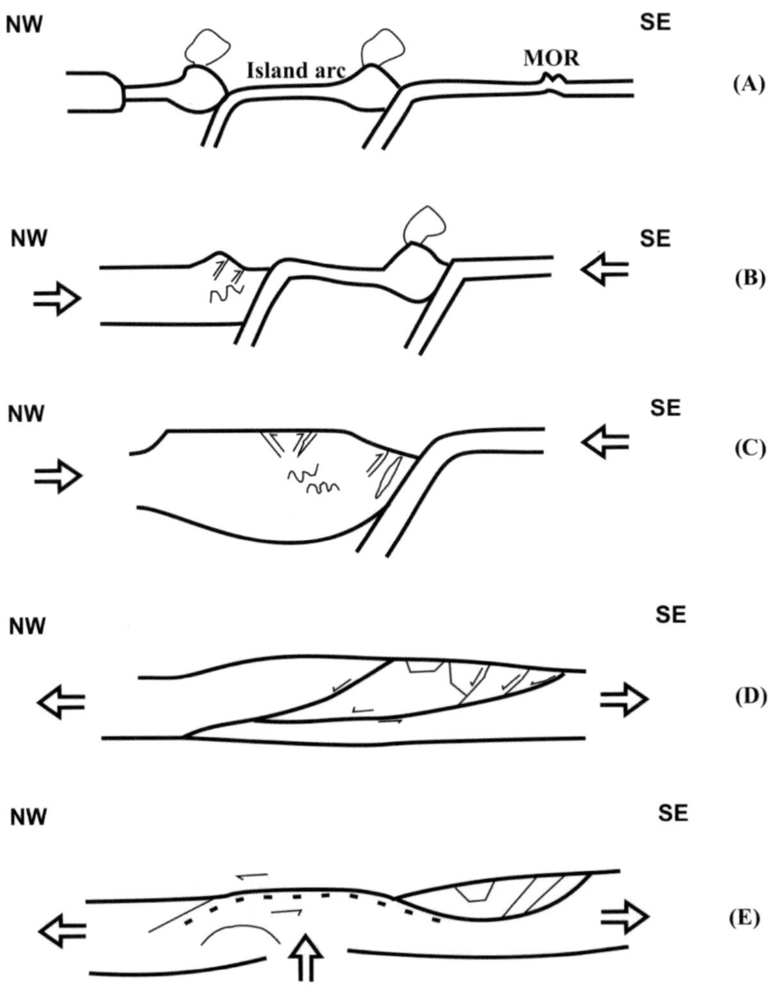

Figure 8.
A cartoon displaying the different stages of the evolution of the Arabian-Nubian Shield after [88].

Gebel Umm Tawat, North Eastern Desert indicates its depositional age as 585 ± 13 Ma [85].

The tectonic evolution of the ANS is commonly divided into three major stages, namely: (1) subduction stage (_870–635 Ma) during which oceanic crust, island arc volcano-sedimentary sequences, and plutonic rocks formed; (2) continental collision (640–650 Ma) resulting from continuing convergence between East and West Gondwana to form the East African orogen [3]; and (3) post-collision stage (580–540 Ma), evidenced by stabilization of ANS crust accompanied by the cutting of a vast peneplain [86]. Development of sedimentary basins and emplacement of increasingly alkaline igneous rocks took place during the last two stages (e.g., [87]). Finally, the ANS stabilized as continental crust by Early Cambrian time (~525 Ma) [1]. The tectonic scenario for ED of Egypt can be summarized as follows [88]:

1. intra-oceanic island arcs were formed in the Mozambique Ocean (**Figure 8A**)

2. suture possibly by a continental block in the western Egypt (**Figure 8B**)

3. arc accretion led to substantial lithospheric thickening (**Figure 8C**). At this stage, conductive heating of the lithospheric root decreased the strength of the crust. The thickened crust became gravitationally unstable and collapsed, which, in turn, led to extension (**Figure 8D**).

4. Crustal thinning, through large low-angle normal shear zones, allowed the intrusion of A-type granites. The isostatic rebound and the intrusion of these granites contributed to the doming of the lower crust and the development of metamorphic core complexes such as the Meatiq domes (**Figure 8E**). Sedimentary basins, bordered by normal faults, were formed at the upper crustal levels as a response to the extension and allowed the deposition of

5. post-orogenic molasse sequences as the Hammamat molasses group.

5. Concluding remarks

1. The Precambrian rocks of Egypt represent the northwestern part of the Arabian-Nubian Shield, which was formed during the Pan-African orogenic cycle (950–450 Ma) [89]. Island arc volcanic rocks and ophiolitic sequences formed between 700 and 800 Ma [6], and then, they were obducted in the earlier stage of the Pan-African orogeny. The Pan-African orogenic event in Egypt ended at about 615 Ma, and subsequent crustal uplifting and extensional collapse occurred within the 610–550 Ma time span [89]. This post-collision stage was characterized by the emplacement of large masses of Dokan volcanics (610–560 Ma) and shallow-level A-type granites (610–550 Ma) [6]. The most common rock units of the ED of Egypt are grouped into an ophiolitic suite/island arc assemblage and post-orogenic intrusions (**Figure 9**) [54]. **Table 1** summarizes the geochemical characteristics, tectonic setting, and age dating of the different rock assemblages.

2. Collectively, the different types of the ED ophiolites fall geochemically and tectonically into two separate groups: MORB-like ophiolites formed in a back-arc tectonic setting and SSZ ophiolites of fore-arc tectonic setting. The tectonic setting of the ophiolites changed from MORB to SSZ with time. Formation of an intra-oceanic island arcs and related volcaniclastic sediments is followed by

the incorporations of ophiolite fragments into the volcaniclastic matrix to form "ophiolitic mélange" through tectonic and/or concurrent sedimentary and tectonic processes to be formed in an interarc or back-arc basin [19]. Subsequent to this stage, the volcanic eruptions of bimodal-evolved island arcs are contemporaneously or shortly followed by deposition of volcaniclastic sediments in an arc-rift basin known as "intra-arc basin" [26].

3. The "gneissic domes" are metamorphic core complexes that were formerly interpreted to have been formed either in a compressional setting or in an extensional regime. However, the obtained age data indicated that the ED granite gneissic and migmatitic rocks are juvenile in origin and Neoproterozoic.

4. Granitoid rocks in the ED include (1) old calcalkaline, I-type, syn- to late-orogenic granitoid assemblages (880–610 Ma) and (2) younger commonly alkaline, post-orogenic to anorogenic granitoid assemblages emplaced between 600 and 475 Ma. Most of the older granitoids and phase I younger granitoids are of I-type character, displaying metaluminous, calcalkaline geochemical characteristic plot in the area of volcanic arc granites (VAG), whereas younger phase granitoids are mainly alkaline, of A-type granites, and of within-plate tectonic setting (WPG). Generally, I-type granitoids were interpreted to result from melting of an amphibolitic crust and dated at approximately 760–650 Ma. The origin of A-type granites is consistent with the melting of a juvenile Neoproterozoic mantle source that assimilated some older crustal materials or as anatectic melts of various crustal sources.

5. Dokan volcanics, sedimentary basins, and post-orogenic A-type granites were interpreted to have been formed in an extensional or rifting regime. This rifting event may have created accommodation space for the Hammamat molasse sediments that accumulated in a structurally controlled intermontane basin.

6. More geochemical and age dating studies are required to characterize the different rock units and to determine their ages, compositional variations, and consequently, to construct the tectonic evolution of the Neoproterozoic crust through time.

Event	Lithology		Age	References
550 Ma				
Post-collision stage		Younger Granites Hammamat sediments Dokhan Volcanics	630-550 Ma	Beyth et al. (1994), Wilde and Youssef (2000), Breitkreuz et al. (2010) , Be'eri-Shlevin et al. (2011), Moghazi et al. (2012),
635 Ma	**Collision of east and west Gondwana (Avigad and Gvirtzman., 2009)**			
Island arc stage	Island arc assemblage	Older granitoids Rhyolite flows Volcanogenic breccia Andesite flows Metasediments ± ***Banded iron formation***	850-635 Ma	Stern and Hedge (1985), Abdel-Rahman & Doig (1987) Ali et al. (2009),
	Ophiolite sequence	Deep marine sediments Pillow and massive basalts Diabase dykes Gabbro Serpentinites	810-730 Ma	Kröner et al. (1992) Zimmer et al. (1995) Andresen et al. (2009) Ali et al. (2010)
900 Ma				

Figure 9.
Lithostratigraphy, major tectonic events, and ages of the basement complex in the Eastern Desert of Egypt from [54].

Rock units	Geochemical characters and tectonic settings	Age dating
Granite gneisses, amphibolites and migmatites	Granite gneiss core varies from enriched REE, to alkaline granite REE pattern; from enriched in HFSE, Rb, Ga, and total REE showing A-type characters and within plate tectonic setting, to granitic gneisses of calcalkaline and of I-type granites and of volcanic arc tectonic setting	631 Ma for Meatiq granite gneisses [10]; 700 Ma for Hafafit granite gneisses [38]; Wadi Beitan migmatitic granitic gneisses yielded 719 ± 10, and 744 ± 10 Ma [2]
Ophiolite assemblage	MORB-like ophiolites formed in a back-arc tectonic setting and SSZ ophiolites of fore-arc tectonic setting; flat REE MORB volcanic ophiolites, and depleted to slightly enriched LREE pattern for volcanic SSZ ophiolites	Wadi Gerf 741 ± 21 Ma [80]; 730–750 Ma [79]; Wadi Ghadir 746 ± 19 Ma, [80]; Fawakhir 736.5 ± 1.2 Ma [10]
Island arc assemblage	Arc metavolcanics with slightly fractionated REE patterns of island arc affinity; bimodal metavolcanics with felsic lavas are slightly LREE-depleted, whereas basalt is slightly LREE-enriched; arc-related volcaniclastic sediments (interarc to intra-arc basin)	Island-arc volcanics to 720–770 Ma [6]; Shadli island arc metavolcanics 712 Ma [28]; 750 Ma for the volcanic and volcano-sedimentary rocks [49]
Granitoid rocks	Old tonalite-granodiorite, calcalkaline, I-type, volcanic arc granites, syn- to late-orogenic granitoids; most of younger granitoids are mainly alkaline, A-type granites, and of within-WPG tectonic setting	The Aswan Tonalite and the monumental granite intruded at 606 Ma, respectively [38]; younger granites covers a time span between 600 and 550 Ma, [6]; A-type granites dated ca. ~600 Ma [81]; 590 ± 3.1 Ma [10] for Um Had younger granite
Dokhan volcanics	Generally, have steep LREE and nearly flat HREE; medium-K to high-K calcalkaline affinity and continental arc to within plate tectonic setting (transitional setting)	592 and 630 Ma for acidic Dokan volcanics [83]
Hammamat molasse sediments	Greywackes with relatively high Zr, Nb, Y, and TH and relatively low Cr, Ni and V, and Sc; have enriched LREE pattern; of active continental margin or continental island arcs tectonic setting and appear to be formed in pull-apart intermontane basins; their sources are felsic source of evolved magmatic island arcs and active continental margin together with minor inputs from calcalkaline island arcs or ophiolitic mafic rocks	Depositional age 585 ± 13 Ma [75, 85]

Table 1.
Summary of geochemical characteristics, tectonic settings, and age dating of the different Neoproterozoic rock assemblages of the ED of Egypt.

Author details

Gaafar A. El Bahariya
Geology Department, Tanta University, Tanta, Egypt

*Address all correspondence to: gbahariya@yahoo.com

References

[1] Robinson FA, Foden JD, Collins AS, Payne JL. Arabian shield magmatic cycles and their relationship with Gondwana assembly: Insights from zircon U-Pb and Hf isotopes. Earth and Planetary Science Letters. 2014;**408**: 207-225

[2] Ali KA, Kröner A, Hegner E, Wong J, Li S-Q, Gahlan HA, et al. U-Pb zircon geochronology and Hf-Nd isotopic systematics of Wadi Beitan granitoid gneisses, South Eastern Desert, Egypt. Gondwana Research. 2015;**27**:811-824

[3] Stern RJ. Arc assembly and continental collision in the Neoproterozoic East African Orogen: Implications for the consolidation of Gondwanaland. Annual Review of Earth and Planetary Sciences. 1994;**22**:319-351

[4] Johnson PR, Woldehaimanot B. Development of the Arabian-nubian shield: Perspectives on accretion and deformation in the northern east African orogeny and the assembly of gondwana. In: Yoshida M, Windley BE, Dasgupta S, editors. Geological Society, London, Special Publications. Vol. 206. 2003. pp. 289-325

[5] Moussa EMM, Stern RJ, Manton WI, Ali KA. SHRIMP zircon dating and Sm/Nd isotopic investigations of Neoproterozoic granitoids, Eastern Desert, Egypt. Precambrian Research. 2008;**160**:341-356

[6] Stern RJ, Hedge CE. Geochronologic constraints on late Precambrian crustal evolution in the Eastern Desert of Egypt. American Journal of Science. 1985;**285**:7-127

[7] Abd El-Rahman Y, Polat A, Dilek Y, Kusky TM, El-Sharkawy M, Said A. Cryogenian ophiolite tectonics and metallogeny of the central Eastern Desert of Egypt. International Geology Review. 2012;**54**:1870-1884

[8] El-Gaby S, List FK, Tehrani R. Geology, evolution and metallogenesis of the Pan-African Belt in Egypt. In: El Gaby S, Greiling RO, editors. The Pan African Belt of Northeast Africa and Adjacent Areas. Braunschweig/ Wiesbaden: Vieweg & Sohn; 1988. pp. 17-68

[9] Bregar M, Bauernhofer A, Pelz K, Kloetzli U, Fritz H, Neumayr P. A late Neoproterozoic magmatic core complex in the Eastern Desert of Egypt: Emplacement of granitoids in a wrench-tectonic setting. Precambrian Research. 2002;**118**:59-82

[10] Andresen A, El-Rus MMA, Myhre PI, Boghdady GY, Corfu F. U-Pb TIMS age constraints on the evolution of the Neoproterozoic Meatiq Gneiss Dome, Eastern Desert, Egypt. International Journal of Earth Sciences. 2009;**98**: 481-497

[11] Greiling RO, Kröner A, El-Ramly MF. Structural interference patterns and their origin in the Pan-African basement of the Southeastern Desert of Egypt. In: Kröner A, Greiling RO, editors. Precambrian Tectonics Illustrated. Stuttgart, Germany: Schweitzerbart'sche Verlagsbuchhandlung; 1984. pp. 401-412

[12] Fritz H, Wallbrecher E, Khudier AA, Abu El Ela F, Dallmeyer RD. Formation of Neoproterozoic metamorphic core complexes during oblique convergence, Eastern Desert, Egypt. Journal of African Earth Sciences. 1996;**23**:311-329

[13] El Bahariya GA, Abu Anbar MM. Geology and tectonometamorphic evolution of the Meatiq dome, Eastern

Desert, Egypt: Pan-African crustal evolution. In: Proc. 1st International Symposium on Geophysics. Tanta, Egypt; 1999. pp. 406-427

[14] El Bahariya GA. Geology and petrology of Neoproterozoic syntectonic anatectic migmatites around Wadi Abu Higlig, Hafafit region, Eastern Desert, Egypt. Egyptian Journal of Geology. 2008;**52**:25-54

[15] Ali KA, Andresen A, Stern RJ, Manton WI, Omar SA, Maurice AE. U-Pb zircon and Sr-Nd-Hf isotopic evidence for a juvenile origin of the c 634 Ma El-Shalul Granite, Central Eastern Desert, Egypt. Geological Magazine. 2012;**149**:783-797

[16] Stern RJ. Neoproterozoic formation and evolution of Eastern Desert continental crust—The importance of the infrastructure-superstructure transition. Journal of African Earth Sciences. 2017:1-13

[17] Ries A, Shackelton R, Graham R, Fitches W. Pan-African structures, ophiolites and melang in the Eastern Desert Egypt: A traverse at 26 N. Journal of the Geological Society of London. 1983;**140**:75-95

[18] Kroner A. Ophiolites and the evolution of tectonic boundaries in the late Proterozoic Arabian-Nubian shield of Northeast Africa and Arabia. Precambrian Research. 1985;**27**:277-300

[19] El Bahariya GA. Classification and origin of the Neoproterozoic ophiolitic mélange of the Central Eastern Desert of Egypt: Evidences for back-arc basin deformation. Tectonophysics. 2012; **568–569**:357-370

[20] El Bahariya GA. Classification of the Neoproterozoic ophiolites of the Central Eastern Desert, Egypt based on field geological characteristics and mode of occurrence. Arabian Journal of Geosciences. 2018;**11**:313

[21] El Bayoumi RM. Ophiolites and mélange complex of Wadi Ghadir area, Eastern Desert, Egypt. Journal of King Abdulaziz University. 1983;**6**:329-342

[22] Akaad MK, Noweir AM, Abu El Ela AM. The volcano-sedimentary association and ophiolites of Wadi Mubarak, Eastern Desert, Egypt. In: Proc. Inter. Conf. G.S.E. Spec. Publ. Vol. 69. 1995. pp. 231-248

[23] Akaad MK, Noweir AM, Abu El Ela AM, El Bahariya GA. The Um Esh olistostromal melange, Qift-Quseir region, central Eastern Desert, Egypt. Journal of Geology. 1997;**41**(1):465

[24] Abdel-Karim AM, Azzaz SM, Moharem AF, El-Alfy HM. Petrological and geochemical studies on the Ophiolite and island arc association of Wadi Hammariya, Central Eastern Desert, Egypt. The Arabian Journal for Science and Engineering. 2008;**33**(1): 117-138

[25] Gahlan HA, Azer MK, Khalil AES. The Neoproterozoic Abu Dahr ophiolite, South Eastern Desert, Egypt: Petrological characteristics and tectonomagmatic evolution. Mineralogy and Petrology. 2015;**109**:611-630

[26] El Bahariya GA. Geology, geochemistry, and source characteristics of Neoproterozoic arc-related clastic metasediments, Central Eastern Desert, Egypt. Arabian Journal of Geosciences. 2018;**11**:87

[27] Maurice AE, Basta FF, Khiamy AA. Neoproterozoic nascent island arc volcanism from the Nubian shield of Egypt: Magma genesis and generation of continental crust in intra-oceanic arcs. Lithos. 2012;**132-133**:1-20

[28] Stern RJ, Kröner A, Rashwan AA. A late Precambrian (~710 Ma) high volcanicity rift in the South Eastern Desert of Egypt. Geologische Rundschau. 1991;**80**:155-170

[29] Akaad MK, Noweir AM. Geology and lithostratigraphy of the Arabian Desert orogenic belt of Egypt between Lat. 25 35 and 26 30 N. In: Evolution and Mineralization of the Arabian Nubian Shield. Bull. 3, 4. Jeddah: Inst. Applied Geol., Univ.; 1980. pp. 127-135

[30] Akaad MK, Noweir AM, Kotb H. Geology and petrochemistry of the granite association of the Arabian Desert Orogenic Belt of Egypt between Lat. 25° 35′ and 26°30′. Delta Journal of Science. 1979;**3**:107-151

[31] Liégeois JP, Stern RJ. Sr-Nd isotopes and geochemistry of granite-neiss complexes from the Meatiq and Hafafit domes, Eastern Desert, Egypt: No evidence for pre-Neoproterozoic crust. Journal of African Earth Sciences. 2010; **57**:31-40

[32] Stern RJ, Gottfried D. Petrogenesis of a late Precambrian (575-600 Ma) bimodal suite in Northeast Africa. Contributions to Mineralogy and Petrology. 1986;**92**:492-501

[33] Abu El Ela AM, El Bahariya GA. The Hammamat molasse sediments between Wadi Umm Esh and Wadi Muweilih, Qift-Quseir region, Central Eastern Desert. Egyptian Journal of Geology. 1997;**41/2A**:5-35

[34] El Bahariya GA, Abu Anbar MM. Geology and petrochemistry of two constracting deformed granite plutons and their enclaves, Central Eastern Desert, Egypt. In: The Second Intern. Conf., on the Geology of Africa, Assuit, Egypt. Vol. 1. 2001. pp. 293-319

[35] Khudeir A, Abu El-Rus M, El-gaby S, El-nady O, Bishara W. Sr-Nd isotopes and geochemistry of the infrastructural rocks in the Meatiq and Hafafit core complexes, Eastern Desert, Egypt: Evidence for involvement of pre-Neoproterozoic crust in the growth of Arabian-Nubian shield. Island Arc. 2008;**17**:90-108

[36] Kröner A, Krüger J, Rashwan AA. Age and tectonic setting of granitoid gneisses in the Eastern Desert of Egypt and South-West Sinai. Geologische Rundschau. 1994;**83**:502-513

[37] Farahat ES, Mohamed HA, Ahmed AF, El Mahallawi MM. Origin of I- and A-type granitoids from the Eastern Desert of Egypt: Implications for crustal growth in the northern Arabian-Nubian shield. Journal of African Earth Sciences. 2007;**49**:43-58

[38] Finger F, Dorr W, Gerdes A, Gharib M, Dawoud M. U-Pb zircon ages and geochemical data for the monumental granite and other granitoid rocks from Aswan, Egypt: Implications for the geological evolution of the Western margin of the Arabian Nubian shield. Mineralogy and Petrology. 2008;**93**: 153-183

[39] Zimmer M, Kroner A, Jochum KP, Reischmann T, Todt W. The Gabal Gerf complex: A Precambrian N-MORB ophiolite in the Nubian shield, NE Africa. Chemical Geology. 1995;**123**: 29-51

[40] Abd El-Rahman Y, Polat A, Dilek Y, Fryer BJ, El-Sharkawy M, Sakran S. Geochemistry and tectonic evolution of the Neoproterozoic incipient arc-fore arc crust in the Fawakhir area, Central Eastern Desert, Egypt. Precambrian Research. 2009;**175**:116-134

[41] Abd El-Rahman Y, Polat A, Dilek Y, Fryer BJ, El-Sharkawy M, Sakran S. Geochemistry and tectonic evolution of the Neoproterozoic Wadi Ghadir ophiolite, Eastern Desert, Egypt. Lithos. 2009;**113**:158-178

[42] El Bahariya GA, Arai S. Petrology and origin of Pan-African serpentinites with particular reference to chromian spinel composition, Eastern Desert, Egypt: Implication for supra-subduction zone ophiolite. In: The Third International Conference on the Geolgy

of Africa, Asssuit, Egypt. Vol. 1. 2003. pp. 371-388

[43] El Bahariya GA. Petrology, mineral chemistry and metamorphism of two pan-African ophiolitic metagabbro occurrences, Central Eastern Desert, Egypt. Egyptian Journal of Geology. 2006;**50**:183-202

[44] El Bahariya GA. Geology, compositional variation and petrogenesis of possible MORB-type ophiolitic massive and pillowed metabasalts from the Pan-African belt, Eastern Desert, Egypt. Egyptian Journal of Geology. 2007;**51**:41-59

[45] El Bahariya GA. Geology, mineral chemistry and petrogenesis of Neoproterozoic metamorphosed ophiolitic ultramafics, Central Eastern Desert, Egypt: Implications for the classification and origin of the ophiolitic mélange. Egyptian Journal of Geology. 2008a;**52**:55-82

[46] Basta FF, Maurice AE, Bakhit BR, Ali KA, Manton WI. Neoproterozoic contaminated MORB of Wadi Ghadir ophiolite, NE Africa: Geochemical and Nd and Sr isotopic constraints. Journal of African Earth Sciences. 2011;**59**: 227-242

[47] Azer MK, Stern RJ. Neoproterozoic (835–720 Ma) serpentinites in the Eastern Desert, Egypt: Fragments of forearc mantle. Journal of Geology. 2007;**115**:457-472

[48] Abdel-Karim A-AM, Ali S, Helmy HM, El-Shafei HM. A fore-arc setting of the Gerf ophiolite, Eastern Desert, Egypt: Evidence from mineral chemistry and geochemistry of ultramafites. Lithos. 2016;**263**:52-65

[49] Ali KA, Robert J, Stern RJ, Mantona WI, Kimurab J, Khameesc HA. Geochemistry, Nd isotopes and U-Pb SHRIMP zircon dating of Neoproterozoic volcanic rocks from the Central Eastern Desert of Egypt: New insights into the ~750 Ma crust-forming event. Precambrian Research. 2009;**171**: 1-22

[50] Ali-Bik MW, Hassan SM, Abou El Maaty MA, Abd El Rahim SH, Abayazeed SD, Abdel Wahab W. The late Neoproterozoic Pan-African low-grade metamorphic ophiolitic and island-arc assemblages at Gebel Zabara area, Central Eastern Desert, Egypt: Petrogenesis and remote sensing—Based geologic mapping. Journal of African Earth Sciences. 2018;**144**:17-40

[51] El Habaak H. Petrogenesis and tectonic implications of the rock succession hosting banded iron formation at Um Anab Area, North Eastern Desert of Egypt. In: The Fourth International Conference on the Geology of Africa, Assiut-Egypt. Vol. 2. 2005. pp. 479-513

[52] Abu Anbar MM, El Bahariya GA. Petrology and geochemistry of metamorphosed bimodal volcanic suite, Meatiq Dome Area, Central Eastern Desert, Egypt. In: Fifth Intern. Conf. on Geochem, Alexandria, Egypt. Part 1. 2001. pp. 119-146

[53] Abu El Ela F. Bimodal volcanism of the Igla Eliswid-Um Khariga metavolcanics, Eastern Desert, Egypt. Journal of African Earth Sciences. 1992; **14**(4):477-491

[54] Khalil I, Khalil K, El-Shazly AE, Lehmann B. Late Neoproterozoic banded iron formation (BIF) in the central Eastern Desert of Egypt: Mineralogical and geochemical implications for the origin of the Gebel El Hadid iron ore deposit. Ore Geology Reviews. 2015;**69**:380-399

[55] Hegde VS, Chavadi VC. Geochemistry of late Archean metagreywackes from the western Dharwar craton, South India: Implications for provenance and nature

of the late Archean crust. Gondwana Research. 2009;**15**(2):178-187

[56] Basta FF, Maurice AE, Bakhit BR, Azer MK, El-Sobky AF. Intrusive rocks of the Wadi Hamad Area, North Eastern Desert, Egypt: Change of magma composition with maturity of Neoproterozoic continental island arc and the role of collisional plutonism in the differentiation of arc crust. Lithos. 2017;**288–289**:248-263

[57] Moghazi AM. Petrology and geochemistry of Pan-African granitoids, Kab Amiri area, Egypt—Implications for tectonomagmatic stages in the Nubian shield evolution. Mineralogy and Petrology. 2002;**75**:41-67

[58] Küster D, Liégeois JP. Sr, Nd isotopes and geochemistry of the Bayuda Desert high-grade metamorphic basement (Sudan): an early Pan-African oceanic convergent margin, not the edge of East Saharan ghost craton? Precambrian Research. 2001;**109**:1-23

[59] Hussein AA, Ali MM, El-Ramly MF. A proposed new classification of the granites of Egypt. Journal of Volcanology and Geothermal Research. 1982;**14**:187-198

[60] Moghazi AM, Mohamed FH, Kanisawa S. Geochemical and petrological evidence of calc-alkaline and A-type magmatism in the Homrit Waggat and El-Yatima areas of eastern Egypt. Journal of African Earth Sciences. 1999;**29**(3):535-549

[61] Pearce JA, Harris NBW, Tindle AG. Trace element discrimination diagrams for the tectonic interpretation of granitic rocks. Journal of Petrology. 1984;**25**:956-983

[62] Eby GN. Chemical subdivision of the A-type granitoids: Petrogenetic and tectonic implications. Geology. 1992;**20**: 641-644

[63] Bonin B. A-type granites and related rocks: Evolution of a concept, problems and prospects. Lithos. 2007;**97**:1-29

[64] Mohamed FH, Moghazi AM, Hassanen MA. Geochemistry, petrogenesis and tectonic setting of late Neoproterozoic Dokhan-type volcanic rocks in the Fatira area, eastern Egypt. International Journal of Earth Sciences. 2000;**88**:764-777

[65] Abdel Wahed AA, Ali KG, Khalil MA, Abdel Gawad AE. Dokhan volcanics of Gabal Monqul area, North Eastern Desert, Egypt: Geochemistry and petrogenesis. Arabian Journal of Geosciences. 2010. DOI: 10.1007/s12517-010-0136-z

[66] Stern RJ, Gottfried D, Hedge CE. Late Precambrian rifting and crustal evolution in the Northeast Desert of Egypt. Geology. 1984;**12**:168-172

[67] Ressetar R, Monard JR. Chemical composition and tectonic setting of the Dokhan volcanic formation, Eastern Desert, Egypt. Journal of African Earth Sciences. 1983;**1**:103-112

[68] Eliwa HA, Kimura JI, Itaya T. Late Neoproterozoic Dokhan volcanics, North Eastern Desert, Egypt: Geochemistry and petrogenesis. Precambrian Research. 2006;**151**:31-52

[69] Pearce JA, Gale GH. Identification of ore-depositional environment from trace-element geochemistry of associated igneous host rocks. In: Volcanic Processes in Ore Genesis. Vol. 7. London: Special Pub. Geol. Soc; 1977. pp. 14-24

[70] Taylor SR, McLennan SM. The Continental Crust: Its Composition and Evolution. Oxford: Blackwell Scientific Publications; 1985. p. 312

[71] Abd El-Rahman Y, Polat A, Fryer BJ, Dilek Y, El-Sharkawy M, Sakran S. The provenance and tectonic setting of the

Neoproterozoic Um Hassa Greywacke Member, Wadi Hammamat area, Egypt: Evidence from petrography and geochemistry. Journal of African Earth Sciences. 2010;**58**:185-196

[72] Willis KM, Stern RJ. Age and geochemistry of late Precambrian sediments of the Hammamat series from the Northeastern Desert of Egypt. Precambrian Research. 1988;**42**:173-187

[73] Bhatia MR. Plate tectonics and geochemical composition of sandstones. Journal of Geology. 1983;**91**:611-627

[74] Bhatia MR, Crook KAW. Trace element characteristics of greywackes and tectonic setting discrimination of sedimentary basins. Contributions to Mineralogy and Petrology. 1986;**92**: 181-193

[75] Roser BP, Korsch RJ. Determination of tectonic setting of sandstone-mudstone suites using SiO_2 content and K_2O/Na_2O ratio. Journal of Geology. 1986;**94**(5):635-650

[76] Grothaus BT, Ehrlich R, Eppler DT. Facies analysis of the Hammamat sediments, Eastern Desert, Egypt. In: 5th Conference on African Geology, Cairo, Annals of the Geological Survey of Egypt. Vol. 9. 1979. pp. 564-590

[77] Fritz H, Messner M. Intramontane basin formation during oblique convergence in the Eastern Desert of Egypt: Magmatically versus tectonically induced subsidence. Tectonophysics. 1999;**315**:145-162

[78] Sturchio NC, Sultan M, Sylvester P, Batiza R, Hedge C, El-Shazly EM, et al. Geology, age, and origin of Meatiq Dome: Implications for the Precambrian stratigraphy and tectonic evolution of the Eastern Desert of Egypt. In: Al-Shanti AM, editor. King Abdulaziz University Faculty of Earth Sciences. Bull, 6. 1984. pp. 127-143

[79] Ali KA, Azer MK, Gahlan HA, Wilde SA, Samuel MD, Stern RJ. Age constraints on the formation and emplacement of neoproterozoic ophiolites along the Allaqi-Heiani suture, South Eastern Desert of Egypt. Gondwana Research. 2010;**18**:583e595

[80] Kröner A, Todt W, Hussein IM, Mansour IM, Mansour M, Rashwan AA. Dating of late Proterozoic ophiolites in Egypt and the Sudan using the single grain zircon evaporation technique. Precambrian Research. 1992;**59**:15-32

[81] Ali KA, Zoheir BA, Stern RJ, Andresen A, Whitehouse MJ, Bishara WW. Lu-Hf and O isotopic compositions on single zircons from the North Eastern Desert of Egypt, Arabian-Nubian shield: Implications for crustal evolution. Gondwana Research. 2016; **32**:181-192

[82] Katzir Y, Eyal M, Litvinovsky BA, Jahn BM, Zanvilevich AN, Valley JW, et al. Petrogenesis of A-type granites and origin of vertical zoning in the Katharina pluton, Gebel Mussa (Mt. Moses) area, Sinai, Egypt. Lithos. 2007; **95**:208-228

[83] Breitkreuz C, Eliwa H, Khalaf I, El Gameel K, Sergeev S, Larionov A. Neoproterozoic SHRIMP U-Pb zircon ages of Dokhan volcanics in the Northern part of the Eastern Desert, Egypt. In: 33rd International Geological Congress Oslo; August 6–14th 2008. Abstract. 2010

[84] Shalaby A, Stu K, Fritz H, Makroum F. The El Mayah molasse basin in the Eastern Desert of Egypt. Journal of African Earth Sciences. 2006;**45**:1-15

[85] Wilde SA, Youssef K. Significance of SHRIMP U-Pb dating of the imperial porphyry and associated Dokhan volcanics, Gebel Dokhan, North Eastern Desert, Egypt. Journal of African Earth Sciences. 2000;**31**:403-413

[86] Avigad D, Sandler A, Kolodner K, Stern RJ, McWilliams M, Miller N, et al. Mass-production of Cambro-Ordovician quartz-rich sandstone as a consequence of chemical weathering of Pan-African terranes: Environmental implications. Earth and Planetary Science Letters. 2005;**240**:818-826

[87] Johnson PR. Post-amalgamation basins of the NE Arabian shield and implications for Neoproterozoic III tectonism in the Northern East African orogen. Precambrian Research. 2003; **123**:321-337

[88] Blasband B, White S, Brooijmans P, De Boorder H, Visser W. Late Proterozoic extensional collapse in the Arabian-Nubian shield. Journal of the Geological Society, London. 2000;**157**: 615-628

[89] Greiling RO, Abdeen MM, Dardir AA, Akhal H, El Ramly MF, Kamal El Din GM, et al. A structural synthesis of the Proterozoic Arabian-Nubian shield in Egypt. Geologische Rundschau. 1994; **83**:484-501

Chapter 2

Tsunami Elemental Signatures in the Samoan Islands: A Case Study

Shaun Williams

Abstract

This study uses Itrax X-ray fluorescence element data available for Samoan sediment cores, obtained from three separate locations after the 2009 tsunami in this region, to identify its elemental characteristics in the coastal landscape. Normalization of data using Al reveals a distinct elevated elemental signature for this event at sites which had experienced inundation. This provides benchmarks for identifying comparable signatures in the core profiles which likely represent past tsunamis to have inundated each site. Such information can support a better understanding of the frequency and longer-term threats posed by tsunamis in this region. The findings presented are consistent with benchmark tsunami Itrax observations at Little Pigeon Bay following the 2016 Kaikōura Tsunami in New Zealand. Furthermore, they reinforce the normalization of Itrax element data using Al to interpret tsunami episodes in sediment cores, in addition to using high resolution core scanning as an effective non-destructive tool to screen likely tsunami deposits for more targeted multi-proxy analysis.

Keywords: sediment cores, tsunami deposits, Itrax core scanner, X-ray fluorescence, element characteristics

1. Introduction

Tsunami deposit studies involve the application and use of a wide range of interdisciplinary techniques including, but not limited to, geology, stratigraphy, geochronology, geochemistry, geophysics, numerical modeling, sedimentology, micro- and macro-paleontology, geography, geomorphology, historical and ethno-historical studies, archeology, statistical and contextual studies [1]. Such studies can support an understanding of the long-term hazard frequency as well as potential magnitude, sources, and risk of tsunamis in coastal areas. These in turn can help to underpin long-term coastal resilience planning in vulnerable locations.

Identifying past tsunami deposits in sedimentary records typically requires multiple characteristic criteria to be met [2]. For example, tsunamis can deposit distinct sedimentary units which can extend up to several kilometers and fine inland and upwards within the deposit. Similarly, distinct elevations in elemental concentrations of sodium, sulfur and chlorine (which are salinity indicators) can be observed in tsunami deposits relative to underlying and overlying sediment units. Deposits are typically preserved in coastal settings such as wetlands, which absorb wave energy and enable sediment to settle out of suspension as the tsunami wanes to normal sea level. The use of trace element, principal component, radiometric and isotope analysis to help distinguish marine from terrestrially sourced sediment and determine the ages of specific sediment units can help to constrain a tsunami layer. A more detailed description of these examples, including a comprehensive list of tsunami deposit characteristics can be found in [1–8].

The use of geochemical proxies to characterize modern tsunami deposits and identify paleo-events has gained widespread use within the last few decades [3]. These include the characterization of elemental and magnetic susceptibility transitions in sedimentary sequences, isotope signatures, electrical conductivity and salinity [3]. In addition, such proxies have also been used to characterize the extent of tsunami inundation in areas which lack sedimentary transitions visible to the naked eye. For example, in [4–8].

A tool which has become more frequently used in tsunami deposit studies is the Itrax core scanner [9, 10]. The Itrax is a non-destructive, multifunction core scanning instrument which produces high-resolution X-ray fluorescence (XRF) elemental and magnetic susceptibility data, as well as optical and radiographic imagery. This enables the detection of very fine-scale elemental transitions in core sequences [9].

Raw Itrax elemental data of sediment cores from highly organic environments are subject to inherent uncertainties associated with organic dilution, matrix and instrumental effects [11–14]. However, organic dilution affects the raw datasets due to the closed-sum effect which in turn, can affect analytical interpretations if it is not adequately accounted for [11]. Normalization against Al can provide an organic-filtered dataset representative of the lithogenic faction of the sediment, which is more reliable for tsunami interpretation [8, 11]. Itrax studies of tsunami deposits prior to the results presented in [8] had not accounted for organic dilution in the datasets, often resulting in ambiguous Itrax interpretations of tsunami episodes.

In this study, available Itrax data for sediment cores collected from the Samoan Islands following the 2009 tsunami in this region [15, 16] (**Figure 1**), are normalized against Al and compared with tsunami Itrax observations at Little Pigeon Bay following the 2016 Kaikōura tsunami in New Zealand [8]. A distinct elevated elemental signature associated with the 2016 event was observed at Little Pigeon Bay, with comparable signatures representing characteristic types of events deeper in the sedimentary record. This provides a basis for reviewing the Samoan datasets to assess whether comparable trends are observed for the 2009 event. Distinct elevations in bromine (Br), chlorine (Cl), sulfur (S), calcium (Ca), iron (Fe), silicon (Si), titanium (Ti), rubidium (Rb) and potassium (K), are particularly targeted as these elements are known to represent marine influences [17, 18]. The findings are discussed with conclusions provided in the context of tsunami deposits and longer-term hazards in this region.

1.1 The 2009 tsunami in the Samoan Islands

The 2009 tsunami in the Samoan Islands deposited a distinct sedimentary unit in much of the coastal areas it inundated [2, 16]. The southern and eastern coastlines of Upolu Island were particularly affected [15]. In most cases, sediment deposited by the event comprised of distinct marine-derived calcareous sand units overlying dark brown and/or loamy pre-event soils. Most of these generally displayed an upwards fining within the unit from coarse sand to mud, and were commonly preserved in inundated coastal wetlands where water stagnation up to several days after the event was observed in some areas. These deposits also displayed distinct elemental elevations compared with underlying pre-event soil units as determined through coarse portable XRF (pXRF) analysis, in addition to marked changes in grain size [19].

This provides a reference for corroborating distinct elevated elemental signatures for this event in this study, and for interpreting potentially older tsunami

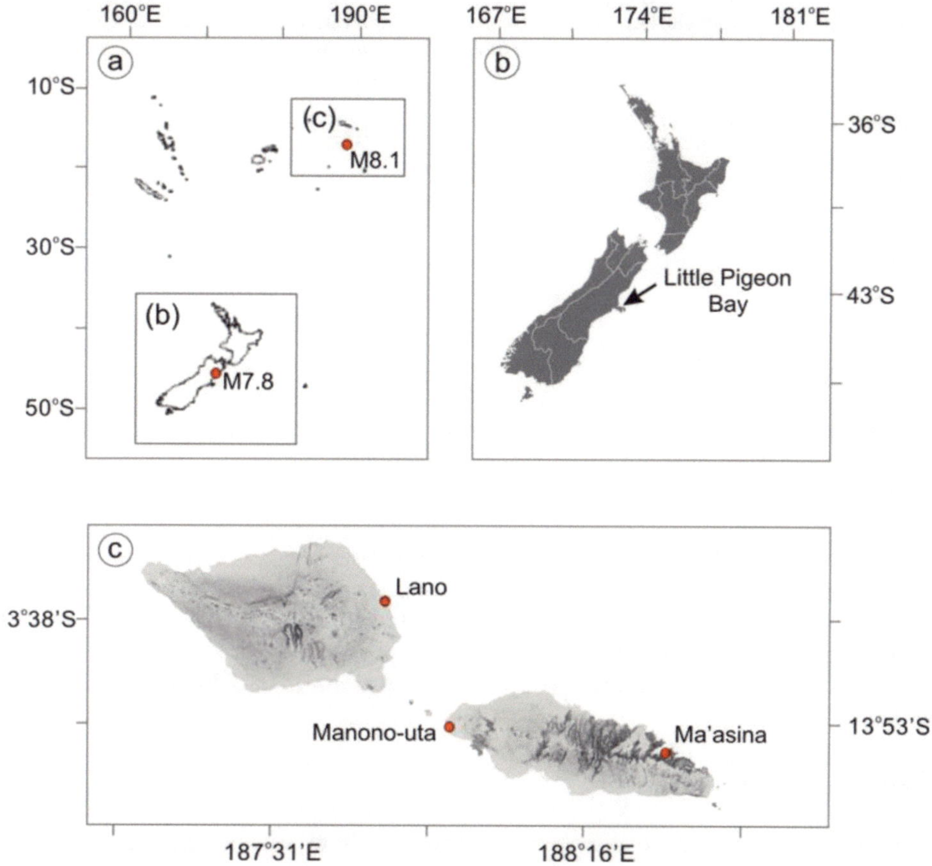

Figure 1.
(a) Location of New Zealand and Samoa in the Southwest Pacific showing the 2016 M7.8 Kaikōura earthquake and 2009 M8.1 Samoa-Tonga earthquake locations. (b) Little Pigeon Bay site, New Zealand. (c) Core site locations in the Samoan Islands. Detailed descriptions of the Little Pigeon Bay site and core profiles are provided in [8], with the Samoan sites and profiles provided in [16, 19].

episodes in the core profiles. For example, an interpreted older episode in a core profile might comprise of a distinct calcareous sand unit bound by an upper and a lower soil unit, and which displays distinct elevated elemental signals relative to the adjacent soils that are comparable to the 2009 tsunami signature.

2. Methods and materials

2.1 Study area and sampling

The data used in this study were obtained from three cores which were sampled in November 2010 using a hand-held D-Corer from three separate locations; (1) 0.7 m core extracted ~20 m inland of the shore at Ma'asina (S1); (2) 2 m core extracted ~75 m inland of the shore at Manono-uta (S2); and (3) 1.5 m core extracted ~150 m inland of the shore at Lano (S3) (**Table 1**).

The cores were sampled from coastal wetlands inland of the shore, and for the case of Manono and Ma'asina, these wetlands were exposed to inundation during the 2009 tsunami. At Ma'asina, a discernible calcareous sand deposit was observed between 0.4–0.8 m depth in the core. The embayment which this area is located has been impacted in the past by far-field tsunamis such as the 1960 Valdivia

Site	Coordinate location	Distance from shore (m)	Elevation above mean sea level (m)	Core length (m)	Sample environment (m)
Lano	13037.176′ S; 172011.938′ W	150	8	1.5	Coastal wetland
Ma'asina	13056.607′ S; 171033.585′ W	40	8	0.7	Coastal wetland
Manono-uta	13052.120′ S; 172004.263′ W	75	3	2.0	Coastal wetland

Table 1.
Core site locations and descriptions used in this study.

tsunami and 1952 tsunamis originating from the Chile/Peru region, and the 1957 tsunami originating in the Aleutian Islands. However, a distinct calcareous sand unit indicative of the 2009 event was not discernable to the naked eye at the surface of the core.

No discernible calcareous sand deposits were observed in the Manono-uta core, though distinct changes in grain size, organic content (loss on ignition) and indicative pXRF elemental compositions comparable with the characteristics of the 2009 tsunami deposits on eastern Upolu were reported in [19]. At Lano, a distinct calcareous sand deposit was observed at ~1 m depth intercalated between dark brown soil units.

2.2 Itrax data and analysis

Itrax XRF involves the excitation of a sample by X-rays using either a molybdenum (Mo) or chromium (Cr) anode X-ray tube, causing the sample to fluoresce. That is, the sample emits secondary X-rays that are distinctive of the elements which they were emitted from. The relative intensity of these fluorescent signals provides an indication of the elemental composition of the sample [9, 10]. The Mo-tube is commonly used in most applications and is more appropriate for detecting heavier elements, while the Cr- tube more adequately detects lighter elements. For example, Al and Si. Nevertheless, an X-ray exposure time of approximately 10–20 s on the sample using a Mo- tube and the Itrax Q-spec procedure is adequate for obtaining acceptable data for these light elements [10].

Itrax elemental data used in this study were initially presented in [19] and were obtained in 2012 using the Itrax core scanner at the Australian Nuclear Science and Technology Organisation (ANSTO). The scanner was fitted with a Mo- tube and a magnetic susceptibility meter, with XRF scans performed on each core at 30 kV and 55 mA using 10 s exposure time and sampled at 500 μm step size. In comparison, Itrax analysis of the Little Pigeon Bay cores were carried out using similar settings at the University of Auckland [8], with a negligible difference of 4 s exposure time between the two datasets [20].

The samples were air dried and the surfaces cleaned to minimize water content and matrix effects in the core prior to analysis. Data quality was ensured through normalization against the Al detected for each core, with profile gaps resolved through interpolation by a moving average curve. This enabled more meaningful trends to be observed in the core sequences. Identified elevated elemental signals relative to adjacent units were compared for consistency with detected changes in unit stratigraphy, grain size, loss on ignition (indicative of organic content if the changes are greater than 5%), and pXRF results for these cores, provided in [19].

2.3 Tsunami deposit identification

Normalized data was used to identify distinct elevated elemental signals in the core profiles comparable to benchmark tsunami signatures provided in [8]. Elevated signals at the surfaces of the Ma'asina and Manono cores are particularly screened for benchmark signatures indicative of the 2009 tsunami. Identified elevated elemental units are compared with descriptions of sites, core stratigraphies, and analogue tsunami deposits of the 2009 tsunami in [16, 19], to interpret distinct tsunami episodes in the core profiles.

3. Results and interpretations

The Samoan cores show comparable elemental profile trends to those observed at Little Pigeon Bay when normalized against Al (**Figure 2**). At Little Pigeon Bay, a sharp sedimentary contact denoted by distinct elevations in elemental levels

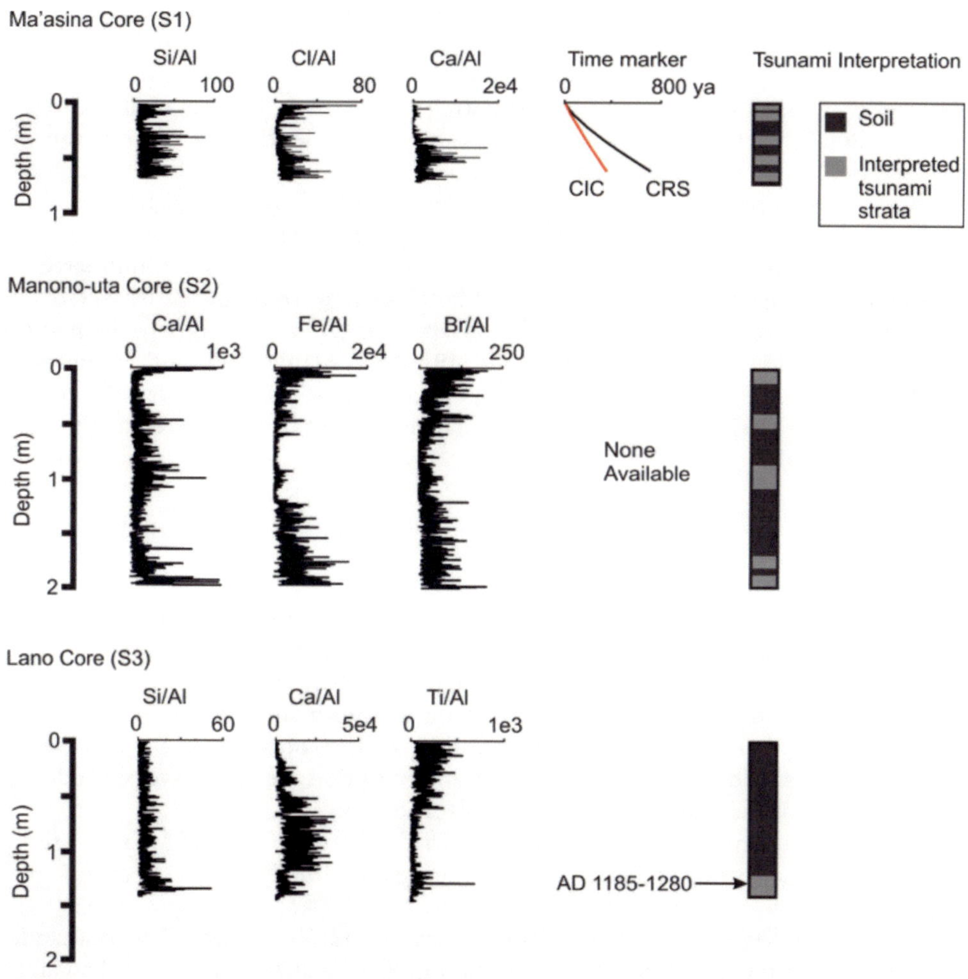

Figure 2.
Selected Itrax data (normalized over Al) for profiles S1 (Ma'asina), S2 (Manono-uta) and S3 (Lano). Time markers shown at Ma'asina are based on ^{210}Pb constant initial concentration (CIC) and constant rate of supply (CRS) ages up to 11 cm profile depth provided in [19] and are forecast using a power function curve. Calibrated ^{14}C age shown at Lano was sourced from [19]. All time markers presented are indicative only and provide tentative benchmarks for interpretation.

associated with the 2016 event directly overlies the pre-inundated sediment surface. This provided a benchmark tsunami signature at this site, with similar characteristic signals observed deeper in the cores [8].

However, significant data gaps are observed in the Samoan profiles, especially the Lano and Manono-uta cores. These gaps are associated with the Al counts approaching detection limit and/or non-detection. In contrast, the non-detection of Al in the Ma'asina core accounted for only 7% of the total dataset compared with 61 and 63% for the Lano and Manono-uta cores, respectively.

At Manono-uta, distinct elevated signals are observed at ~0.06 m depth for most of the detected elements except Si. The Si trend exhibits a relatively low and stable background elemental signal with no discernable changes in the profile, indicating that offshore and terrestrial sources for Si are not abundant at this site. Manono-uta was inundated by the 2009 tsunami with flow depths up to 2 m in some areas, including the core site in this study, with the elevated signals at the core surface likely representing the elemental signature of this event. Characteristic signals deeper in the profile which appear lower in magnitude than the 2009 signature are likely representing: (1) smaller events; (2) other types of coastal processes (e.g., storm surge or rainfall-related flooding); and/or (3) artifacts associated with interpolation and matrix effects.

Comparable trends are observed at Ma'asina where a distinct elevated signal occurs at ~0.05 m. This site was also inundated during the 2009 tsunami which suggests that the elevated signal is representing this event. ^{210}Pb ages available for this core in [19] suggests that similar signals observed at depth could be representing historical tsunamis such as the 1960 Valdivia tsunami, or storm surges which are known to have impacts this area (such as the 1991 Tropical Cyclone Val).

At Lano, elevated elemental signals are not apparent at the surface of the core. Compared with the sites at Manono-uta and Ma'asina, this site was not inundated during the 2009 tsunami [15], thus a signal for this event would not be expected. Interestingly, a strongly elevated signal is observed at 1.4 m which is consistent with a distinct calcareous sand unit described in [19]. An available calibrated ^{14}C age of AD 1185–1280 at 1.39 m obtained from plant fragments within this unit [19], indicates that a potentially significant inundation event might have occurred during, or after, this period.

4. Discussion

Normalization of data against Al in the Samoan cores reveals similar characteristic profile trends analogous to the findings in the Little Pigeon Bay study. This includes the identification of a distinct elemental signal of the 2009 tsunami event at Manono and Ma'asina. While data gaps associated with detection limits and low Al count rates are present in each of the profiles, the interpolated plots provide an adequate means to identify distinct elemental signals that are indicative of sudden and/or distinct influences in the coastal landscape.

The 2009 tsunami benchmark signals identified at Manono-uta and Ma'asina are unique in that they were not clearly detected during initial analysis of these cores in [19]. The characteristic signals deeper in the core at Ma'asina probably represent historical tsunamis that are known to, or may have, inundated this site. Potential candidates include the far-field 1960 Valdivia Tsunami, 1957 Aleutian Tsunami, and possibly the local 1917 Samoa-Tonga and far-field 1868 Arica Tsunamis [19] (**Figure 3**).

The lack of benchmark elemental signatures for storm surges in these Islands makes it difficult to associate any of the signals detected in this study with tropical cyclones. Furthermore, the distance of each core site from the shore makes it more likely that the signals detected deeper in the cores are representing older and/or

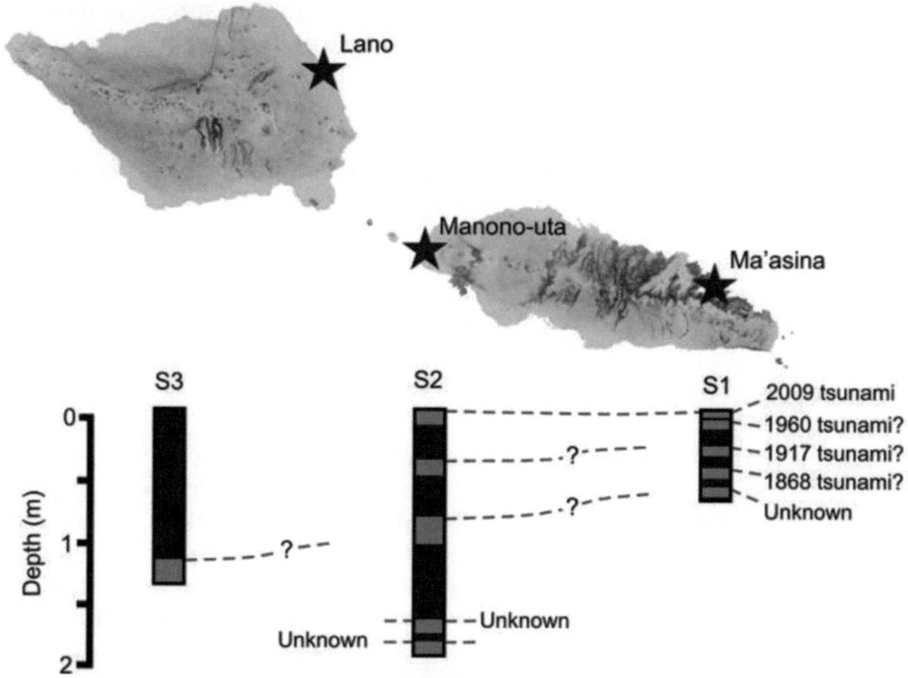

Figure 3.
Interpreted tsunami associated strata in the Samoan cores based on elemental signatures and available geochronological time markers.

unknown tsunami events. However, it is possible that smaller signals than the 2009 benchmark could be representing storm surge inundation.

At Manono-uta, tsunamis to severely inundate this site and leave a signature in the landscape are more likely to be associated with events approaching from southerly directions. It is therefore probable that the characteristic, though weaker, signal observed at 0.45 m depth is representing the local 1917 event, which may have had similar impacts to the 2009 event [21]. However, a definitive correlation is difficult without chronological time-markers at this site.

The relative depth of the AD 1185–1280 (or younger) potential event, denoted by the strong signal at 1.4 m depth in the Lano core, compared with characteristic signals at 1.9 m in the Manono-uta profiles, suggests a possible event association. The depth which these signals occur implies that they are older than the deepest signal exhibited at Ma'asina. However, despite the short core length obtained from Ma'asina, extrapolating the range of available ^{210}Pb time-markers for this site in [19] suggests that the deepest signal could represent an inundation event up to several centuries ago. Furthermore, the signals detected at Ma'asina likely corroborate this embayment providing favorable conditions for preserving evidence of both local and far-field tsunamis to impact this region.

While potentially major regionally and far-field sourced transpacific paleotsunamis have been suggested in the literature [22, 23], a distinct association between such events with the signals detected in the Samoan cores is ambiguous given the limited time markers available. Nevertheless, it is probable that if a major transpacific tsunami sourced at the Tonga-Kermadec zone had impacted the Samoan Islands, it would have left a distinct elemental signal in the landscape similar to, or stronger than, the 2009 signature. The occurrence of strong elemental signals at depth in Manono-uta and Lano suggest that these sites experienced a significant inundation event at some point in the past. Whether these are representing the same event and whether they signify the extent of a major transpacific or a more localized event similar to the 2009 and 1917 tsunamis, is unclear and requires more detailed investigation.

Importantly, due to non-detection of Al and subsequent data gaps, the Lano and Manono-uta sites are only providing true representations of 39 and 37% of their total analyzed profiles, respectively, compared with 93% for the Ma'asina site. While interpolation enables more meaningful elemental profiles to be produced, it is likely that fine-scale signals at Lano and Manono-uta are diluted and not adequately represented. Consequently, the interpretations presented for these sites should be used with care.

5. Conclusion

This study set out to reassess available Itrax elemental datasets for the Samoan Islands to determine whether an elemental signature associated with the 2009 tsunami in this region is observed. After normalization of the Itrax data against Al, a distinct elemental signal for this event is identified at two of the three core sites reassessed, which provided benchmarks for identifying older potential events at each site. The dearth of evidence for tropical cyclone-related elemental benchmarks in these islands, particularly >20 m inland of the shore, limits the association of any of the detected signals with potential storm deposits. Future investigations involving isotope geochemistry and principal component analysis of likely sources would help to clarify this.

The results presented in this study are consistent with tsunami Itrax observations in the Little Pigeon Bay study, and highlights the importance of normalizing Itrax data using Al in tsunami deposit investigations. This enables a more accurate representation of tsunami elemental signatures in core profiles, and for the case presented in this study, enables an interpretive improvement of tsunami episodes compared with previous studies on these cores. The significant data gaps in the Lano and Manono-uta elemental profiles associated with low Al detection, limits their use in accurately interpreting older potential events at these sites.

Nevertheless, the findings in this study reinforce the use of high resolution Itrax elemental data as an effective tool for characterizing and identifying tsunami inundations in the coastal landscape. Furthermore, they demonstrate the potential for elemental signatures to be used as a rapid screening tool to target more detailed multi-proxy investigations of identified signals, including their potential sources and longer-term hazard implications.

Acknowledgements

This research was supported by NIWA through New Zealand government strategic science investment funding (SSIF), projects CARH1902 and PRAS1901. Itrax data for Samoa and ^{210}Pb dates reproduced in this study were obtained in 2012 through the Australian Institute for Nuclear Sciences and Engineering (AINSE) Grant 12/119. The author acknowledges Tim Davies, Catherine Chagué, Atun Zawadzki, Patricia Gadd, James Goff, Ilyas Qasim, Johnny Ah Kau, Faigame Sale and Mulipola Ausetalia Titimaea for their contributions to earlier associated aspects of this study.

Conflict of interest

None.

Notes

Readers should also refer to citations contained within the references listed below for additional details, where relevant, on various aspects presented in this chapter.

Author details

Shaun Williams
National Institute of Water and Atmospheric Research (NIWA), Christchurch, New Zealand

*Address all correspondence to: shaun.williams@niwa.co.nz

References

[1] Goff J, Chagué-Goff C, Nichol SL, Jaffe B, Dominey-Howes D. Progress in palaeotsunami research. Sedimentary Geology. 2012;**243-244**:70-88. DOI: 10.1016/j.sedgeo.2011.11.002

[2] Chagué-Goff C, Schneider J-L, Goff JR, Dominey-Howes D, Strotz L. Expanding the proxy toolkit to help identify past events—Lessons from the 2004 Indian Ocean Tsunami and the 2009 South Pacific Tsunami. Earth-Science Reviews. 2011;**107**:107-122. DOI: 10.1016/j.earscirev.2011.03.007

[3] Chagué-Goff C, Szczuciński W, Shinozaki T. Applications of geochemistry in tsunami research: A review. Earth-Science Reviews. 2017;**165**:203-244. DOI: 10.1016/j.earscirev.2016.12.003

[4] Chagué-Goff C, Goff J, Wong HKY, Cisternas M. Insights from geochemistry and diatoms to characterise a tsunami's deposit and maximum inundation limit. Marine Geology. 2015;**359**:22-34. DOI: 10.1016/j.margeo.2014.11.009

[5] Chagué C, Sugawara D, Goto K, Goff J, Dudley W, Gadd P. Geological evidence and sediment transport modelling for the 1946 and 1960 tsunamis in Shinmachi, Hilo, Hawaii. Sedimentary Geology. 2018;**364**:319-333. DOI: 10.1016/j.sedgeo.2017.09.010

[6] Judd K, Chagué-Goff C, Goff J, Gadd P, Zawadzki A, Fierro D. Multi-proxy evidence for small historical tsunamis leaving little or no sedimentary record. Marine Geology. 2017;**385**:204-215. DOI: 10.1016/j.margeo.2017.01.002

[7] Kitamura A, Ito M, Sakai S, Yokoyama Y, Miyairi Y. Identification of tsunami deposits using a combination of radiometric dating and oxygen-isotope profiles of articulated bivalves. Marine Geology. 2018;**403**:57-61. DOI: 10.1016/j.margeo.2018.04.003

[8] Williams SP, Zhang T, Chague-Goff C, Williams J, Goff J, Lane EM, et al. Sedimentary and geochemical signature of the 2016 Kaikōura Tsunami at Little Pigeon Bay: A depositional benchmark for the banks peninsula region, New Zealand. Sedimentary Geology. 2018;**369**:60-70. DOI: 10.1016/j.sedgeo.2018.03.013

[9] Croudace I, Rindby A, Rothwell RITRAX. Description and evaluation of a new multi-function X-ray core scanner. In: Rothwell R, editor. New Techniques in Sediment Core Analysis. Vol. 267. London, Special Publication: Geological Society; 2006. pp. 51-63

[10] Croudace IW, Rothwell G, editors. Micro-XRF studies of sediment cores: Applications of a non-destructive tool for the environmental sciences. In: Developments in Paleoenvironmental Research. Vol. 17. Dordrecht: Springer; 2015. 656 p. DOI: 10.1007/978-94-017-9849-5

[11] Löwemark L, Chen H-F, Yang T-N, Kylander M, Yu E-F, Hsu Y-W, et al. Normalizing XRF-scanner data: A cautionary note on the interpretation of high-resolution records from organic-rich lakes. Journal of Asian Earth Science. 2011;**40**:1250-1256. DOI: 10.1016/j.seaes.2010.06.002

[12] Longman J, Veres D, Wennrich V. Utilisation of XRF core scanning on peat and other highly organic sediments. Quaternary International. 2018. DOI: 10.1016/j.quaint.2018.10.015

[13] Bloemsma M, Croudace I, Daly JS, Edwards RJ, Francus P, Galloway JM, et al. Practical guidelines and recent advances in the Itrax XRF core scanning procedure. Quaternary International. 2018. DOI: 10.1016/j.quaint.2018.10.044

[14] Jones AF, Turner JN, Daly JS, Francus P, Edwards RJ. Signal-to-noise

ratios, instrument parameters and repeatability of Itrax XRF core scan measurements of floodplain sediments. Quaternary International. 2018. DOI: 10.1016/j.quaint.2018.09.006

[15] Okal EA, Fritz HM, Synolakis CE, Borrero JC, Weiss R, Lynett PJ, et al. Field survey of the Samoa Tsunami of 29 September 2009. Seismological Research Letters. 2010;**81**:577-591. DOI: 10.1785/gssrl.81.4.577

[16] Williams S, Goff J, Ah Kau J, Sale F, Chagué-Goff C, Davies T. Geological investigation of palaeotsunamis in the Samoan Islands: Interim report and research directions. Science of Tsunami Hazards. 2013;**32**(3):156-175. ISSN: 8755-6839

[17] Chagué-Goff C, Chan JCH, Goff J, Gadd P. Late Holocene record of environmental changes, cyclones and tsunamis in a coastal lake, Mangaia, Cook Islands. Island Arc. 2016;**25**: 333-349. DOI: 10.1111/iar.12153

[18] Stewart H, Bradwell T, Bullard J, Davies SJ, Golledge N, McCulloch RD. 8000 years of North Atlantic storminess reconstructed from a Scottish peat record: Implications for Holocene atmospheric circulation patterns in Western Europe. Journal of Quaternary Science. 2017;**32**:1075-1084. DOI: 10.1002/jqs.2983

[19] Williams S. Tsunami hazard, palaeotsunami investigation, numerical modeling and risk implications [thesis]. Samoan Islands: University of Canterbury; 2014. Available from: http://hdl.handle.net/10092/9664

[20] Huang JJ, Löwemark L, Chang Q, Lin TY, Chen HF, Song SR, et al. Choosing optimal exposure times for XRF core-scanning: Suggestions based on the analysis of geological reference materials. Geochemistry, Geophysics, Geosystems. 2016;**17**:1558-1566. DOI: 10.1002/2016GC006256

[21] Okal E, Borrero J, Chagué-Goff C. Tsunamigenic predecessors to the 2009 Samoa earthquake. Earth-Science Reviews. 2011;**107**:127-140. DOI: 10.1016/j.earscirev.2010.12.007

[22] Goff J, McFadgen B, Chagué-Goff C, Nichol SL. Palaeotsunamis and their influence on Polynesian settlement. The Holocene. 2012;**22**:1067-1069. DOI: 10.1177/0959683612437873

[23] Goff J, Goto K, Chagué C, Watanabe M, Gadd PS, King DN. New Zealand's most easterly palaeotsunami deposit confirms evidence for major trans-Pacific event. Marine Geology. 2018;**404**:158-173. DOI: 10.1016/j.margeo.2018.08.001

Chapter 3

Soil Carbon Biogeochemistry in Arid and Semiarid Forests

Wei-Yu Shi, Xiao-Cong Zhu, Feng-Bao Zhang, Kai-Bo Wang, Lei Deng and Ming-Guo Ma

Abstract

Soil is the largest carbon pool in the terrestrial ecosystem. Even small changes in the soil carbon pool would have huge impacts on atmospheric CO_2 concentrations and thus mitigate or intensify global warming. Global forest contains $383 \pm 30 \times 10^{15}$ g carbon stock in soils to a 1-m depth, which is approximately 50% of the carbon stored in the atmosphere. Arid and semiarid areas with more than 30% of the world's land surface are characterized by low and sporadic moisture availability and sparse or discontinuous vegetation, both spatially and temporally. Vegetation, water, and nutrients are intimately coupled in the semiarid environments with strong feedbacks and interactions occurring across fine to coarse scales. In this chapter, we will review the cutting-edge work in forest soil carbon biogeochemistry undertaken in the last three decades. We also attempt to synthesize recent advances in soil carbon biogeochemistry in arid and semiarid regions and discuss future research needs and directions.

Keywords: carbon cycle, soil respiration, water, soil, ecological restoration

1. Introduction

Arid and semiarid regions cover >30% of the earth's land surface in the world [1]. The soil carbon (C) in arid and semiarid forests plays an important role in global carbon storage and alleviates the increase of atmospheric carbon dioxide (CO_2) concentration, and its contribution to the global carbon cycle is increasingly significant [2]. Afforestation has occurred globally within the framework of the Kyoto Protocol [3] and has the potential to mitigate the rising atmospheric CO_2 concentration caused by anthropogenic emissions [4]. Recent studies also suggested that arid and semiarid ecosystems have strong soil C sequestration potential [5].

Therefore, soil C change in arid and semiarid forest ecosystem is a key process for understanding the global C cycle, assessing the responses of terrestrial ecosystems to climate change and to aid policy makers in making land use/management decisions [6]. In the past, the study of soil carbon cycle was mainly focused on soil organic carbon (SOC), and the behavior of soil inorganic carbon (SIC) was rarely considered. Global "Missing Carbon Sink" reaches $2–3 \times 10^{15}$ g C, and carbonate and carbon fixation in the arid and semiarid regions account for about 1/3 of global "carbon sinking" [7]. The SIC storage in arid and semiarid regions was huge, and it needs more attention in the soil carbon cycle. In this chapter, we will review the cutting-edge work in forest soil carbon biogeochemistry undertaken in the last three decades. We also attempt to synthesize recent advances in soil carbon biogeochemistry in arid and semiarid regions and discuss future research needs and directions.

2. Materials and methods

Traditional measurements of soil physical and chemical properties require the following steps [8, 9]. First, the typical plots are selected in the study area, clipping the vegetation to ground level, and litter (dead plant material) is cleared before soil sampling in each plot. A global positioning system (GPS) is used to determine grid point latitude, longitude, and altitude. Second, Soil samples from different soil layers are collected using a soil drilling sampler. Soil bulk density (D_b) (g cm^{-3}) is assessed by collecting undisturbed soil in a stainless steel cutting ring (volume: 100 cm^3), drying it at 105°C, and weighing it, with three replicates in each plot. Third, the samples from the same layer were mixed to produce one sample in a plot. All soil samples are taken to the laboratory, air-dried, and passed through a 2-mm sieve, and roots and other debris are removed by hand for soil physicochemical analysis. Several methods exist for determining SOC, and wet combustion methods, including Walkley-Black, Mebius, and Colorimetric determination, as well as dry combustion methods, such as elemental and gravimetric analysis, are usually used. Each method has its own advantages and limitations, and all methods require more than three replicates [10]. The soil total carbon content is measured by dry combustion, and the SIC content is calculated by the difference between soil total carbon and SOC content.

Recent studies employing laser-induced breakdown spectroscopy (LIBS) and visible-near infrared diffuse reflectance spectroscopy (vis-NIRS) indicate their potential for rapid in situ soil carbon (SOC and SIC) determination, and these spectroscopic methods differ fundamentally, with LIBS being foremost an elemental analyzer and vis-NIRS a molecular technique. These technologies currently require ideal control conditions, and soil in situ measurement accuracy cannot be confirmed. It is standard practice to pretreat soils using various combinations of air-drying, powdering, sieving, and pelletizing under pressure prior LIBS and vis-NIRS for soil carbon determination in laboratory conditions [11].

3. Factors affecting soil carbon dynamics

3.1 Effect of temperature and precipitation on soil carbon

Climate appeared to strongly modify the effects of afforestation on ecosystem carbon stocks in the arid and semiarid regions [12], due to their effects on the quantity and quality of organic residue soil inputs and on the rates of soil organic matter mineralization and litter decomposition [13, 14]. As water was the major factor that limited plant growth in these regions [15], the SOC accumulation after afforestation was found to vary according to the precipitation level. In regions with precipitation, Zhang et al. estimated the changes in SOC stocks after afforestation of arid and semiarid regions using meta-analysis based on the dataset compiled from published studies [16]. SOC increased in regions with precipitation of 0–250 mm, 250–400 mm, and >400 mm by 54.1, 75.75, and 7.02%, respectively. Jackson et al. found a clear negative relationship between precipitation and changes in SOC stocks after afforestation [17]. The above two cases suggest that the rate of SOC accumulation would decrease with the increase of precipitation, and regions with precipitation of 250–400 mm are ideal for SOC accumulation when afforestation is in arid and semiarid areas. Liu et al. report that soil organic carbon density (SOCD) was significantly higher in areas where the precipitation was greater than 500 mm than where it was less than 500 mm in the Loess Plateau region in China [14]. However, soil erosion could reduce the positive effect of increased precipitation on SOCD in these areas [18]. Semiarid areas are more likely to be cultivated than arid areas, and the semiarid areas are more susceptible to soil erosion by water than the arid areas, so that recently tilled bare soils are exposed to the erosive power of rainfall [14].

The differences in temperatures play a significant role in SOC accumulation processes in ecosystems [19]. Some studies have confirmed that the combination of warmer temperatures and wetter conditions could lead to higher biomass productivity and greater SOC accumulation [14]. Zhang et al. reported that SOC increased by 64.15% in regions with temperatures of 7–15°C, but it increased less than 10% in regions with temperatures of <7.5°C [16]. Relatively higher SOC accumulation in areas with temperatures of <7.5°C could be attributed to the less carbon accumulated in plant biomass; the input of soil organic matter also will be less due to the lower temperatures, as well as the drier conditions; microbial activity is also less intense at lower temperatures; and organic matter is not decomposed rapidly. However, high temperature does not necessarily increase SOC accumulation. Although heat and high precipitation contribute to high net primary productivity (NPP) and high carbon accumulation in plant biomass in tropical regions, climatic conditions also stimulate decomposition and thus reduce SOC stocks [16, 20]. These results suggest that in arid and semiarid regions, 7–15°C is a better option for accumulating C in comparison to <7.5°C and >15°C.

3.2 Effects of soil properties on soil carbon

The effects on soil C and soil properties are important to understand not only because these are often master variables determining soil fertility but also because of the role of soils as a source or sink for C on a global scale [21]. Brahim et al.'s study to develop two models of SOC under clayey and sandy soils in semiarid Mediterranean zones based on physical and chemical soil properties and structural equation modeling (SEM) was adopted to quantify the relative importance of potential direct and indirect pathways in soil properties' effect on SOC [22]. SEM is included in the class of generalized linear models. As a flexible multivariate analysis method that includes factor and path analyses, SEM is useful for evaluating the relative importance of the pathways in hypothetical models and for comparing models with experimental data [23, 24]. For modeling SOC, soil databases composed of various information for organic matter (OM), organic carbon (OC), total nitrogen, pH, D_b, clay, silt (fine and coarse fraction), sand (fine and coarse fraction), and calcium carbonate (C_aCO_3) were used.

"Physical properties" and "chemical properties and D_b/chemical properties" are the latent variables for two types of soils (clayey and sandy soils), and the latent variable is measured by multiple observed variables (i.e., clay, C-silt, F-sand, pH, OM, N, D_b, and OC) (**Figure 1**). Red double arrow line indicates correlations between the measurement errors for observable indicators of the exogenous latent variables. Brahim et al. attributed this fact to the OM and mineral fraction that constitute an organo-mineral complex [22], which are generally associated with clay [25], D_b is associated at a coarse soil fraction as the sand [26]. Brahim et al. also found that in clayey soils, chemical properties and bulk density play the most important role in controlling OC content [22]. The pH, OM, N, and D_b represent the key variables responsible for OC storage. In addition, in sandy soils, the findings show that chemical factors (i.e., OM and pH) are better indicators of OC content than did physical properties. **Figure 1** shows that for clayey and sandy soil model, chemical properties

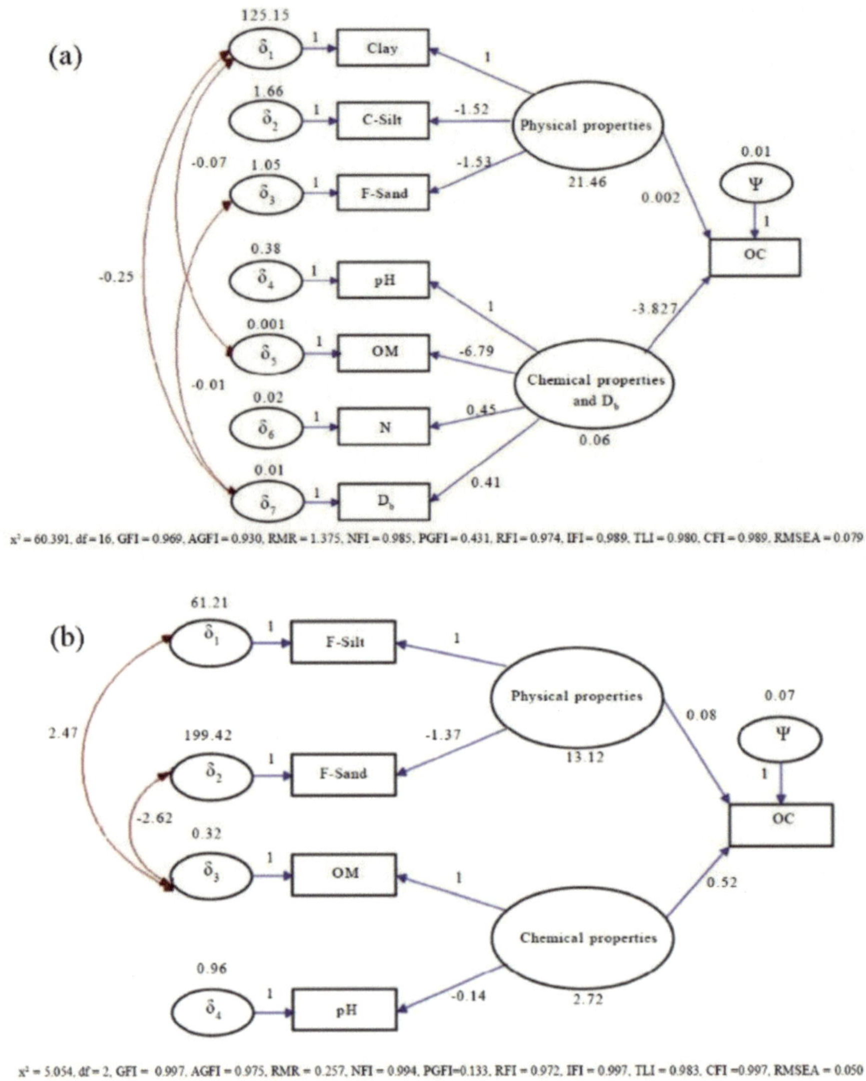

Figure 1.
The estimated parameters of the model predicting SOC in clayey soils (a) and sandy soils (b), respectively, cited from Brahim et al. [22].

and D_b/chemical properties had a stronger effect on OC, than did physical properties, and goodness-of-fit indices for the SEM are all acceptable. We should note the independent effects on OC content between physical properties and chemical properties. The above case studies were conducted mainly in semiarid Mediterranean regions. It is easy to speculate that the conventional relationship between OC and influencing factor in soil is influenced on a global scale, such as that found commonly in northwestern China, western America, and Midwestern Australia. Land degradation and desertification are pervasive in arid and semiarid climate, lands are especially threatened by erosion phenomena, and the restoration of these regions needs afforestation, which inhibits these land degradation phenomena and enhances soil carbon sequestration and soil fertility. Korkanç's study also concluded that afforestation increased the SOC budget, and this situation improved some soil properties, such as increasing water holding capacity (WHC) and total porosity (TP) and reducing D_b and dispersion ratio (DR) over a period of 15 years [27].

3.3 Effect of elevated CO_2 on soil carbon

The changes in the amount of carbon sequestered by soils are closely related to the increase or decrease in the amount of CO_2 accumulation in the atmosphere. Elevated atmospheric CO_2 frequently increases plant production and concomitant soil C inputs, which may cause additional soil C sequestration [28]. While the processes of C sequestration are ultimately regulated at the molecular level, atmospheric CO_2 concentration can greatly affect the way in which terrestrial ecosystems sequester C [29]. Niklaus et al. reported that the increases in leaf litter production at elevated CO_2 may exceed the response in standing biomass [30]. In addition, elevated CO_2 may also induce greater C fluxes from the growing plants to the soil through increasing rates of leaf litter and root material deposition [31]. Thus, elevated atmospheric CO_2 will likely affect soil carbon cycle through its indirect impact on photosynthesis. If C input into the soil is increased, and given that elevated atmospheric CO_2 increases plant production and allocation of photosynthate to below ground components, soil carbon sequestration would be expected to increase [32].

Diaz et al. found that increased C inputs under elevated CO_2 stimulated competition between the soil microbial biomass and plants for soil N, leading to a decline in soil N availability [33]. Hu et al. suggested that elevated CO_2 reduces the amount of N available to microbes through enhanced plant growth [34]. This could result in enhanced C accumulation in grassland soils at elevated CO_2. However, it remains unclear how initial increases in soil C input under elevated CO_2 affect microbial N transformation processes [28].

4. Dynamic characteristics of forest soil carbon

4.1 Evolution characteristics of soil carbon after afforestation

Land degradation and desertification are pervasive in arid and semiarid regions, often resulting in emission of CO_2 into the atmosphere as well as other environmental degradation [35]. Afforestation can increase sequestration of atmospheric carbon dioxide and hence attenuate global warming [36]. Farmland reclamation will exacerbate land degradation and desertification in arid and semiarid regions due to the special climate. Afforestation is the conversion of degraded farmland into vegetation in these regions and renovation without involving natural vegetation. This is similar to the method of Grain for Green Program (GGP) in central and western China [37]. In addition, we consider the choice of tree species to adapt to the arid climate is necessary. The contribution of afforestation to the C cycle has been estimated by many studies on a regional and global scale [6, 38]. Land use and land-cover changes have attracted increasing scientific interest in the past decades in relation to their contribution to potential impacts on soil carbon sequestration and soil nitrogen [39]. Liu et al. estimated the changes in SOC (a) and total nitrogen (TN) stocks (b) after afforestation of arid and semiarid regions using meta-analysis based on the dataset compiled from published studies (**Figure 2**) [40].

Afforestation on different land uses showed different impacts on SOC stock and TN stock. On average across all studies, afforestation significantly increased SOC stock and TN stock by 131 and 88%, respectively. SOC stock and TN stock decreased with different land uses in the following order: BF > CF > GF (CF: afforestation on cropland; GF: afforestation on grassland; and BF: afforestation on barren land); and they also reported significant increases in SOC stock as afforestation was observed for all tree species. SOC and TN accumulations in plantations with different tree species decreased in the following order: broadleaf deciduous > conifer > broadleaf

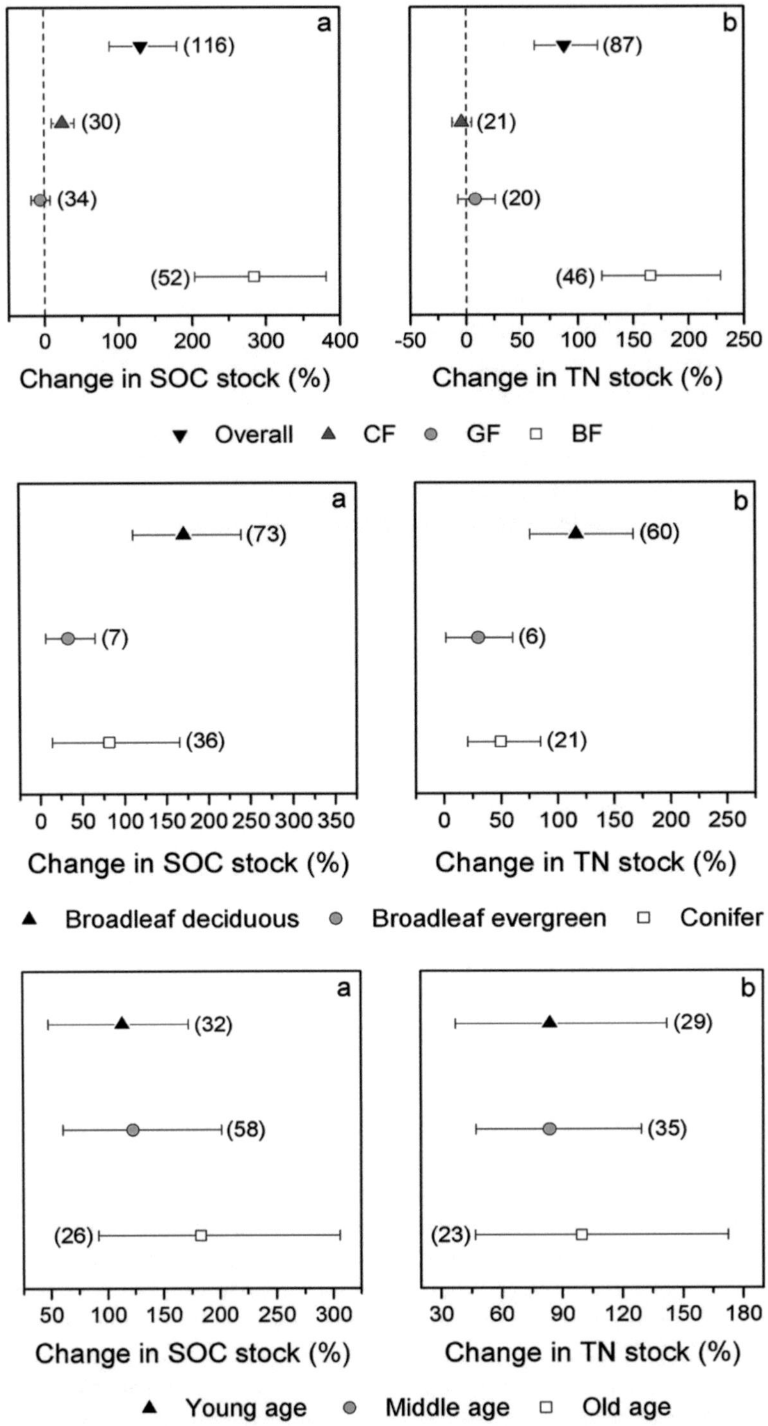

Figure 2.
Changes in soil organic carbon (SOC) (a) and total nitrogen (TN) stocks (b) after afforestation as influenced by prior land-use type, planted tree species, and plantation age, respectively, cited from Liu et al. [40]. CF, afforestation on cropland; GF, afforestation on grassland; and BF, afforestation on barren land. Planted tree species were classified into three categories: broadleaf deciduous, broadleaf evergreen, and conifer. The ages of afforestation were divided into three groups: young age (≤10 year), middle age (>10, ≤30 year), and old age (>30 year).

evergreen. SOC stock significantly increased following afforestation from 114 to 183% with the increase in plantation age. Nonetheless, afforestation-induced changes in SOC stock did not differ significantly among plantation ages. Afforestation significantly increased TN stock by 84–100% for plantations with different ages, with the largest increase being found in plantation with old age. However, the differences in changes in TN stock among plantation ages were not significant [40].

These results suggest that in arid and semiarid regions, BF is a better option for accumulating C and N in comparison to CF, while GF is not recommended as a way to sequester C and N into soils. Korkanç's study showed that the 0–10 cm soil layer of lands afforested with Cedar, a coniferous tree, sequestrated more organic carbon than Black Pine in the central Anatolia region [27]. The inconsistency between the above two research conclusions also proves the necessity to evaluate afforestation efforts using different species of trees on semiarid degraded land as measured by soil SOC and selected soil properties. This also may be useful for determining which species of trees to plant in future afforestation efforts aimed at combating the impacts of global warming [27]. Although all estimates of soil C loss due to land degradation are speculative, the numbers are large (20–30 × 10^{15} g) [35]. Cole et al. through desertification control and adoption of recommended land use and soil management practices, this would amount to 12–20 × 10^{15} g over a 50-year period [41]. After afforestation, SOC and TN accumulations generally showed increasing trends with the increase of plantation age, and restoration age is an important factor to consider when estimating SOC stock and TN stock after afforestation in arid and semiarid regions. Korkanç also reports that SOC values of the afforested lands are generally higher than those in the bare land soils, and the highest SOC value was obtained from the 0 to 10 cm layer in the soils of the Cedar site (1.49%), and the lowest value was from the 10 to 20 cm soil layer in the bare land (0.44%) [27]. According to Lima et al., afforestation of degraded grasslands led to a rise in SOC accumulation in the semiarid regions for a period of 30 years [42].

4.2 Soil carbon cycle process

The soil carbon (C) pool includes organic carbon pools and inorganic carbon pools with carbon stocks of 1555 × 10^{15} and 1750 × 10^{15} g, respectively [43]. Inorganic carbon mainly refers to carbonate carbon existing in arid and semiarid soil. Carbonate can retain atmospheric CO_2 during the formation process, and its formation and turnover have an important impact on the carbon cycle in arid and semiarid regions [2, 44]. Soil carbon cycle mechanisms in arid and semiarid regions include atmospheric pressure transport, carbonate dissolution, and soil water-in-gas percolation [45]. Li et al. showed that the evaporation in semiarid areas is greater than precipitation, forming an oasis landscape dominated by saline-alkali soils. Saline-alkali soil absorbs CO_2 in the air at a slow rate, and the absorbed CO_2 enters the underground saline layer; thus it is a huge potential inorganic carbon sink in the world (**Figure 3**) [46]. The SIC pool affects the SOC pool by affecting the status of soil aggregates, microbial activity, soil pH, and decomposition rate of organic matter. SOC is a very complex continuous mixture of residues of plants, animals, and microorganisms at all stages of decomposition. Many organic compounds in soil are closely related to inorganic soil particles [47].

Soil respiration consists of respiration by plant roots and respiration from catabolism by heterotrophy, mainly by soil microbes. Soil respiration is one of the major processes controlling the carbon budget of terrestrial ecosystems [48], the main export route of SOC and an important source of atmospheric CO_2. Its dynamic changes will directly affect the global carbon balance [49]. Soil temperature is an

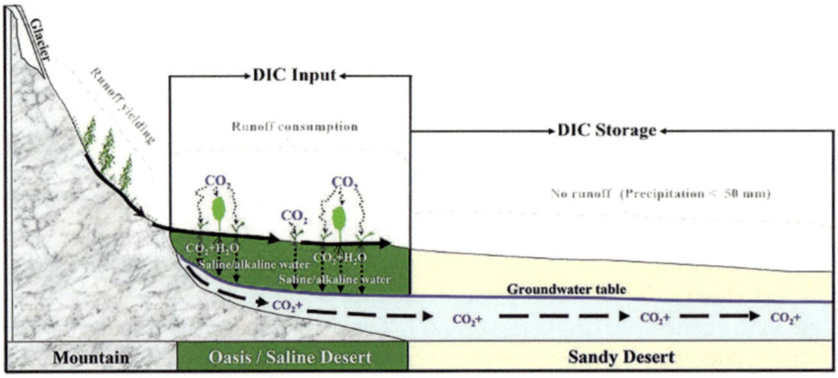

Figure 3.
Schematic diagram of DIC (dissolved inorganic carbon) leaching and transport in a closed arid basin: Tarim Basin, as an example cited from Li et al. [46].

important environmental factor, controlling a complex series of biochemical processes in soil respiration soil respiration rate sensitive to changes in soil temperature, soil temperature change will cause significant changes in soil respiration of terrestrial ecosystems carbon budget patterns have a significant impact [50]. Fierer et al. study have shown that with the increase of soil temperature, soil respiration rate of growth slowed, reducing sensitivity to temperature change, at lower temperatures, soil respiration mainly controlled by temperature changes [51]; when the temperature is high, soil respiration mainly affected by soil moisture and other factors. Soil moisture is a key limiting factor on soil respiration, soil moisture content in most ecosystems and soil respiration was significantly positively related to increased soil moisture will promote soil respiration [52]. Sponseller's study has shown that an increase in soil moisture accelerates the rate of soil respiration by affecting the vegetation's root metabolism and soil microbial activity [53]. After the soil temperature and moisture increase exceed a certain threshold range, microbial activity and soil permeability become lower, which will significantly inhibit soil respiration [54].

5. Main conclusions and future research lines

In the past few decades, many studies have explored the evolution of forest soil carbon after afforestation, but there is still no unified conclusion. We still need three issues in future studies of forest soil carbon biogeochemistry in arid and semiarid regions. First, soil N dynamics and C-N interactions should be focused on for considering soil C accumulation. Second, we should note the effect of changes in soil properties on soil carbon after afforestation in soil carbon cycle, and we need to consider the ecological benefit. Third, we should identify key environmental factors in soil CO_2 sequestration and its influence in climate change.

Acknowledgements

This research has been supported by projects from Chongqing Basic Research and Frontier Exploration Program (cstc2018jcyjAX0187), the Fundamental Research Funds for the Central Universities (SWU116087 and XDJK2017B013), the New Star Foundation on Shaanxi Province Youth Science and Technology (2016KJXX-89), and National Natural Science Foundation of China (Nos. 41201258, 41411140035, and 41671511).

Author details

Wei-Yu Shi[1*†], Xiao-Cong Zhu[1†], Feng-Bao Zhang[2], Kai-Bo Wang[2], Lei Deng[2] and Ming-Guo Ma[1]

1 Research Base of Karst Eco-Environments at Nanchuan in Chongqing of Ministry of Nature Resources and Chongqing Engineering Research Center for Remote Sensing Big Data, School of Geographical Sciences, Southwest University, Chongqing, China

2 State Key Laboratory of Soil Erosion and Dryland Farming on Loess Plateau, Institute of Soil and Water Conservation, Northwest A&F University, Yangling, Shaanxi, China

*Address all correspondence to: shiweiyu@swu.edu.cn

† These authors contributed equally to this chapter.

References

[1] Saco P, Willgoose G, Hancock G. Ecogeomorphology of banded vegetation patterns in arid and semi-arid regions. Hydrology and Earth System Sciences Discussions. 2007;**11**:1717-1730

[2] Lagacherie P, Baret F, Feret JB, Netto JM, Robbez-Masson JM. Estimation of soil clay and calcium carbonate using laboratory, field and airborne hyperspectral measurements. Remote Sensing of Environment. 2008;**112**:825-835

[3] Laganiere J, Angers DA, Pare D. Carbon accumulation in agricultural soils after afforestation: A meta-analysis. Global Change Biology. 2010;**16**:439-453

[4] Nilsson S, Schopfhauser W. The carbon-sequestration potential of a global afforestation program. Climatic Change. 1995;**30**:267-293

[5] Li C, Zhang C, Luo G, Chen X, Maisupova B, Madaminov AA, et al. Carbon stock and its responses to climate change in Central Asia. Global Change Biology. 2015;**21**:1951-1967

[6] Shi J, Cui L. Soil carbon change and its affecting factors following afforestation in China. Landscape and Urban Planning. 2010;**98**:75-85

[7] Karim A, Veizer J, Barth J. Net ecosystem production in the Great Lakes basin and its implications for the North American missing carbon sink: A hydrologic and stable isotope approach. Global and Planetary Change. 2008;**61**:15-27

[8] Liu Y, Dang ZQ, Tian FP, Wang D, Wu GL. Soil organic carbon and inorganic carbon accumulation along a 30-year grassland restoration chronosequence in semi-arid regions (China). Land Degradation and Development. 2017;**28**:189-198

[9] Liu X, Zhang W, Wu M, Ye Y, Wang K, Li D. Changes in soil nitrogen stocks following vegetation restoration in a typical karst catchment. Land Degradation and Development. 2019;**30**:60-72

[10] Sato JH, Figueiredo CC, Marchão RL, Madari BE, Benedito LEC, Busato JG, et al. Methods of soil organic carbon determination in Brazilian savannah soils. Scientia Agricola. 2014;**71**:302-308

[11] Bricklemyer RS, Brown DJ, Turk PJ, Clegg S. Comparing vis–NIRS, LIBS, and combined vis–NIRS-LIBS for intact soil core soil carbon measurement. Soil Science Society of America Journal. 2018;**82**:1482-1496

[12] Xiong D, Shi P, Zhang X, Zou CB. Effects of grazing exclusion on carbon sequestration and plant diversity in grasslands of China—A meta-analysis. Ecological Engineering. 2016;**94**:647-655

[13] Quideau SA, Chadwick OA, Benesi A, Graham RC, Anderson MA. A direct link between forest vegetation type and soil organic matter composition. Geoderma. 2001;**104**:41-60

[14] Liu Z, Shao M, Wang Y. Effect of environmental factors on regional soil organic carbon stocks across the Loess Plateau region. China, Agriculture, Ecosystems and Environment. 2011;**142**:184-194

[15] St. Clair SB, Sudderth EA, Fischer ML, Torn MS, Stuart SA, Salve R, et al. Soil drying and nitrogen availability modulate carbon and water exchange over a range of annual precipitation totals and grassland vegetation types. Global Change Biology. 2010;**15**:3018-3030

[16] Zhang YQ, Liu JB, Jia X, Qin SG. Soil organic carbon accumulation in arid

and semiarid areas after afforestation: A meta-analysis. Polish Journal of Environmental Studies. 2013;**22**:611-620

[17] Jackson RB, Banner JL, Jobbágy EG, Pockman WT, Wall DH. Ecosystem carbon loss with woody plant invasion of grasslands. Nature. 2002;**418**:623-626

[18] Lal R. Soil carbon sequestration to mitigate climate change. Geoderma. 2004;**123**:1-22

[19] Dalias P, Anderson JM, Bottner P, Coûteaux MM. Temperature responses of carbon mineralization in conifer forest soils from different regional climates incubated under standard laboratory conditions. Global Change Biology. 2010;**7**:181-192

[20] Lal R. Forest soils and carbon sequestration. Forest Ecology and Management. 2005;**220**:242-258

[21] Johnson DW, Curtis PS. Effects of forest management on soil C and N storage: Meta analysis. Forest Ecology and Management. 2001;**140**:227-238

[22] Brahim N, Blavet D, Gallali T, Bernoux M. Application of structural equation modeling for assessing relationships between organic carbon and soil properties in semiarid Mediterranean region. International Journal of Environmental Science and Technology. 2011;**8**:305-320

[23] Hou E, Chen C, Kuang Y, Zhang Y, Heenan M, Wen D. A structural equation model analysis of phosphorus transformations in global unfertilized and uncultivated soils: P transformations in global soils. Global Biogeochemical Cycles. 2016;**30**:1300-1309

[24] Gama-Rodrigues AC, Sales MVS, Silva PSD, Comerford NB, Cropper WP, Gama-Rodrigues EF. An exploratory analysis of phosphorus transformations in tropical soils using structural equation modeling. Biogeochemistry. 2014;**118**:453-469

[25] Bayer C, Martinneto L, Mielniczuk J, Dieckow J, Amado TJ. C and N stocks and the role of molecular recalcitrance and organomineral interaction in stabilizing soil organic matter in a subtropical Acrisol managed under no-tillage. Geoderma. 2006;**133**:258-268

[26] Benites VM, Machado PLOA, Fidalgo ECC, Coelho MR, Madari BE. Pedotransfer functions for estimating soil bulk density from existing soil survey reports in Brazil. Geoderma. 2007;**139**:90-97

[27] Korkanç SY. Effects of afforestation on soil organic carbon and other soil properties. Catena. 2014;**123**:62-69

[28] Graaff MAD, Kessel CV, Six J. The impact of long-term elevated CO_2 on C and N retention in stable SOM pools. Plant and Soil. 2008;**303**:311-321

[29] Jones MB, Donnelly A. Carbon sequestration in temperate grassland ecosystems and the influence of management, climate and elevated CO_2. New Phytologist. 2010;**164**:423-439

[30] Niklaus PA, Wohlfender M, Siegwolf R, Körner C. Effects of six years atmospheric CO_2 enrichment on plant, soil, and soil microbial C of a calcareous grassland. Plant and Soil. 2001;**233**:189-202

[31] Allard V, Newton PCD, Lieffering M, Soussana JF, Carran RA, Matthew C. Increased quantity and quality of coarse soil organic matter fraction at elevated CO_2 in a grazed grassland are a consequence of enhanced root growth rate and turnover. Plant and Soil. 2005;**276**:49-60

[32] Xie Z, Cadisch G, Edwards G, Baggs EM, Blum H. Carbon dynamics in a temperate grassland soil after 9

years exposure to elevated CO_2 (Swiss FACE). Soil Biology and Biochemistry. 2005;**37**:1387-1395

[33] Diaz S, Grime JP, Harris J, Mcpherson E. Evidence of a feedback mechanism limiting plant response to elevated carbon dioxide. Nature. 1993;**364**:616

[34] Hu S, Chapin FS, Firestone MK, Field CB, Chiariello NR. Nitrogen limitation of microbial decomposition in a grassland under elevated CO_2. Nature. 2001;**409**:188-191

[35] Lal R. Carbon sequestration in dryland ecosystems. Environmental Management. 2004;**33**:528-544

[36] Bonan GB. Forests and climate change: Forcings, feedbacks, and the climate benefits of forests. Science. 2008;**320**:1444-1449

[37] Feng Z, Yang Y, Zhang Y, Zhang P, Li Y. Grain-for-green policy and its impacts on grain supply in West China. Land Use Policy. 2005;**22**:301-312

[38] Fang J, Chen A, Peng C, Zhao S, Ci L. Changes in forest biomass carbon storage in China between 1949 and 1998. Science. 2001;**292**:2320-2322

[39] Bárcena TG, Kiær LP, Vesterdal L, Stefánsdóttir HM, Gundersen P, Sigurdsson BD. Soil carbon stock change following afforestation in Northern Europe: A meta-analysis. Global Change Biology. 2014;**20**:2393-2405

[40] Liu X, Yang T, Wang Q, Huang F, Li L. Dynamics of soil carbon and nitrogen stocks after afforestation in arid and semi-arid regions: A meta-analysis. Science of the Total Environment. 2017;**618**:1658-1664

[41] Cole V, Cerri C, Minami K, Mosier A, Rosenberg N, Sauerbeck D, et al. Agricultural options for mitigation of greenhouse gas emissions. Climate Change. 1995;**54**:745-771

[42] Lima AMN, Silva IR, Neves JCL, Novais RF, Barros NF, Mendonça ES, et al. Soil organic carbon dynamics following afforestation of degraded pastures with eucalyptus in Southeastern Brazil. Forest Ecology and Management. 2006;**235**:219-231

[43] Lal R, Kimble JM, Follett RF, Stewart BA. Soil processes and the carbon cycle. In: Soil Processes and the Carbon Cycle. CRC press; 1998

[44] Schimel DS, House JI, Hibbard KA, Bousquet P, Ciais P, Peylin P, et al. Recent patterns and mechanisms of carbon exchange by terrestrial ecosystems. Nature. 2001;**414**:169-172

[45] Rey A. Mind the gap: Non-biological processes contributing to soil CO_2 efflux. Global Change Biology. 2015;**21**:1752-1761

[46] Li Y, Wang YG, Houghton RA, Tang LS. Hidden carbon sink beneath desert. Geophysical Research Letters. 2015;**42**:5880-5887

[47] Post WM, Kwon KC. Soil carbon sequestration and land-use change: Processes and potential. Global Change Biology. 2000;**6**:317-327

[48] Shi WY, Tateno R, Zhang JG, Wang YL, Yamanaka N, Du S. Response of soil respiration to precipitation during the dry season in two typical forest stands in the forest–grassland transition zone of the: Loess Plateau. Agricultural and Forest Meteorology. 2011;**151**:854-863

[49] Luo Y, Zhou X, Soil Respiration and the Environment. Elsevier; 2006

[50] Hashimoto S, Carvalhais N, Ito A, Migliavacca M, Nishina K, Reichstein M. Global spatiotemporal distribution

of soil respiration modeled using a global database. Biogeosciences. 2015;**12**(13):4331-4364

[51] Fierer N, Craine JM, McLauchlan K, Schimel JP. Litter quality and the temperature sensitivity of decomposition. Ecology. 2005;**86**:320-326

[52] Conant RT, Dalla-Betta P, Klopatek CC, Klopatek JM. Controls on soil respiration in semiarid soils. Soil Biology and Biochemistry. 2004;**36**:945-951

[53] Sponseller R. Precipitation pulses and soil CO_2 flux in a Sonoran Desert ecosystem. Global Change Biology. 2010;**13**:426-436

[54] Fa KY, Liu JB, Zhang YQ, Wu B, Qin SG, Feng W, et al. CO_2 absorption of sandy soil induced by rainfall pulses in a desert ecosystem. Hydrological Processes. 2015;**29**:2043-2051

Chapter 4

The Geochemical Data Imaging and Application in Geoscience: Taking the Northern Daxinganling Metallogenic Belt as an Example

Jiang Chen, Zhaoxia Mao, Yan Zhao, Chunpeng Zhang and Shan Jiang

Abstract

Geochemical data were predominantly expressed by vector format, the research on geochemical data visualization, i.e., raster data format, was not paid proper attention. A total of 39 geochemical elements in 1:200,000 regional geochemical exploration data were rasterized to form images, and then a geochemical image database was generated. This article has carried out the study on geochemical imaging within Daxinganling metallogenic belt. The metallogenic belt had once carried out the regional geochemical survey, the sampling density was 1 site/4 km^2, and 39 geochemistry elements including the microelement and trace element have been analyzed. Quintic polynomial method was used to implement the geochemical data interpolation, and the cell size of formed geochemical elemental image is 1 km. The images of the geochemical elements were processed by image enhancement methods, and then hyperspectral remote sensing data processing method was used for prospecting target selection, lithology mapping, and so on. The interpreted results have been verified in practice. All the abovementioned suggested a good development prospect for the rasterized geochemical images. Finally the author puts forward using rasterize geochemical images in combination with other geological, geophysical, and remote sensing data to make better use of the geochemical data and be more extensively applied in the geoscience.

Keywords: rasterization, vector, hyperspectrum, geoscience, geochemical spectral

1. Introduction

Geochemical data are typically reported as compositions, in the form of such proportions as weight percents, parts per million, etc., subject to a constant sum (e.g., 100%, 1,000,000 ppm). As an important source of geo-information, geochemical data recording multiple element concentration have been successfully processed by advanced multivariate analytical methods (e.g., factor analysis, cluster analysis, etc.) to identify geological bodies and delineate mineralization-favored space [1–6]. The results of these geochemical data were mainly expressed by vector format, including the colorful geochemical map.

The raster image application in geology was further improved with the development of remote sensing technology. With respect to the application of remote sensing in geology, several books on the geological structural interpretation were published [7, 8]. As the multispectral and hyperspectral imaging rapidly grows, most of the researches paid much attention to the extraction of altered mineral information which were often related to different types of ore deposits [9–16]. These ore deposits include Carlin-type deposit, Archean massive sulfide deposit, skarn-type deposit, and volcanic massive sulfide deposit. Some studies also focused on lithology mapping with hyperspectral tools [17–22].

Only little geochemical data was rasterized. It is partly because the rastering process is more complex, and also the formed raster image could not produce good visual effect due to the low sample density. It is worth mentioning that the geochemical data with vector format can provide relatively simple results; thus the rasterized image appears superfluous. A small amount of research focused on geochemical data rasterization. A technique of metal content on maps was developed [23]. Utilizing ALKEMIA software, Gustavsson et al. [24] designed an interpolation and smoothing method to generate maps including dot maps, color maps, and shaded relief maps.

In this study, geochemical data of the northern Daxinganling metallogenic belt were taken as the experiment area.

A geochemical survey with the scale of 1:200,000 was conducted in a large area of the Daxinganling region [6]. In follow-up to that research, the objectives of the present study are to evaluate the rasterization method of geochemical data obtained from the northern Daxinganling region, use rasterized geochemical data to assist in geological mapping and prospecting target selection, and propose an application of rasterized geochemical data.

2. Experimental area and method

2.1 Experimental area

The northern Daxinganling metallogenic belt was chosen as an experimental area. The Daxinganling metallogenic belt generally refers to an area including the main ridge of the Daxinganling Mountains and both of its east and west slopes.

2.2 Geological background

The Daxinganling region generally refers to an area that includes the main ridge of the Daxinganling Mountains and the eastern and western slopes of the ridge (**Figure 1**). The region is located between the Siberian and North China plates [26, 27]. Several of its tectonic units had been divided (**Figure 1**). In this region, the Proterozoic strata are comprised of epicontinental detritus from volcanic eruptions and carbonate sedimentary formations. The strata constitute the Precambrian crystalline basement [26]. The cap rock is composed of the Paleozoic group, including the Ordovician, Silurian, Devonian, and Carboniferous series, which are sets of epicontinental clastic rock, specifically carbonate rocks interlayered with rock from volcanic eruptions and sedimentary formations. The Mesozoic, Jurassic, and Cretaceous series are primarily comprised of rock from nonmarine volcanic eruptions and sedimentary formations. An important area of concentrated mineralization exists in the western region of the Hulunhu-Eerguna fault (fault ① in **Figure 1**). In this area, many deposits exist [28]. The mineralization is mainly subvolcanic-hydrothermal-type deposit and porphyry deposit. The porphyry deposit is predominantly comprised

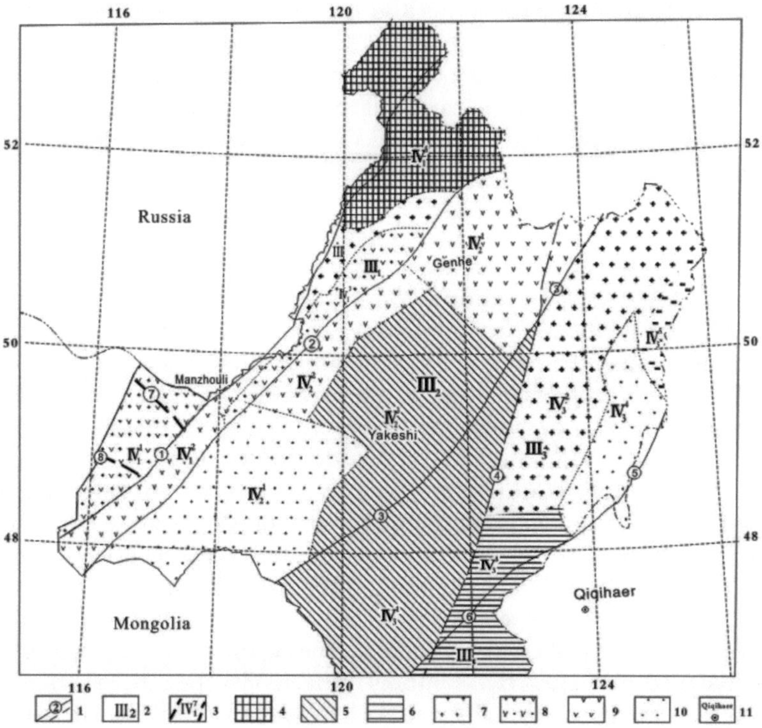

Figure 1.
Geotectonic units of northern Daxinganling metallogenic belt (after [25]). 1, main fault and borderline of the third-order geotectonic unit; 2, the number of third-order geotectonic unit; 3, the number of fourth-order geotectonic unit; 4, Proterozoic lift; 5, early Paleozoic lift; 6, late Paleozoic lift; 7, magmatic area mainly activated in Hercynian period; 8, lift of Mesozoic volcanic and intrusive rocks; 9, basin of Mesozoic volcanic rocks; 10, Cretaceous fault-depression basin; 11, town. The name of main fault: ①, Hulunhu-Eerguna fault; ②, De'erbugan fault; ③, Elunchun-Toudaoqiao fault; ④, Daxinganling main ridge fault; ⑤, Nenjiang fault; ⑥, Arongqi fault; ⑦, Muhaer fault; ⑧, Haligou fault.

of substantial deposits of Wunugetushan porphyry copper, as well as Jiawula and Chaganbulagen lead, zinc, and silver [25].

2.3 Geochemical data

The northern Daxinganling metallogenic belt has two major geographic landscapes, i.e., forest swamp area and semidesert area. The sampling media differed in the two landscapes [29]. The northern Daxinganling metallogenic belt has two major geographic landscapes, i.e., forest swamp area and semidesert area. The sampling media differed in the two landscapes [29]. The 1:200,000 geochemical survey was completed in the Manzhouli area, which covers a total of 13 geological sheets (e.g., Toudaolvdian (M-50-(24)), Manzhoulishi (M-50-(22)), etc.). The geochemical survey was based on stream sediment [30]. The average sampling density was one site per 4 km^2. Sampled material was passed through a 40-mesh sieve before being sent for analysis in the laboratory [31]. The contents of a total of 39 elements, of which seven were major elements, were analyzed, specifically Al, Ca, Fe, K, Mg, Na, Si, and the 32 trace elements (Ag, As, Au, B, Be, Ba, Bi, Cd, Co, Cr, Cu, F, Hg, La, Li, Mn, Mo, Nb, Ni, P, Pb, Sb, Sn, Sr, Th, Ti, U, V, W, Y, Zn, and Zr) [29]. Because 39 kinds of elements are painstakingly picked out, many elements are quite representative. From the periodic table of the elements shown in **Figure 2**, it can be seen that every family has at least one representative element except family 16 and 18.

2.4 Geochemical data rasterizing method

The remote sensing software can be used for the point rastering; most of the data processing and interpretation can be accomplished. In this study, the whole process was completed through using the ENVI software (V. 4.4, Research System Inc., Boulder, CO, USA).

2.4.1 Conversion of data format

The geochemical data obtained from the Daxinganling Mountains were stored in Microsoft XLS format, and 39 elements (oxide) data was included. The arrangement of single elements' data is based on the seriation of the first letters and then followed by oxide. The whole arrangement order is as follows: Ag, As, Au, B, Ba, Be, Bi, Cd, Co, Cr, Cu, F, Hg, La, Li, Mn, Mo, Nb, Ni, P, Pb, Sb, Sn, Sr, Th, Ti, U, V, W, Y, Zn, Zr, Al_2O_3, CaO, Fe_2O_3, K_2O, MgO, Na_2O, and SiO_2. The oxides content were expressed in percentage; the unit of Au and Ag value is ppb. The units of other content values all were ppm. The geographic position was expressed in the format of geographic coordinates.

The file needs to be checked in the sorting way, to inspect whether unqualified data exists. The unqualified data site must be eliminated to ensure quality of the data. Because ENVI cannot directly recognize the file with Microsoft XLS format, the XLS files need to be transformed into TXT format. The latitude and longitude of coordinates were assigned at the first two columns respectively; the other elements were listed afterwards. In Microsoft Excel, the file was saved as TXT format. Additionally, it needs to be noted that if the data content of the geochemical exploration sampling sites is too large even exceeding the permission of Microsoft Excel software, then respectively they need to be else saved in other software, e.g., software Surfer 8 (Golden Software Inc., Golden, USA), to obtain the format that can be recognized by ENVI software. The latter case is suitable for the aeromagnetic data, in which data volume is enormous and usually exceeds the row range of Microsoft Excel software.

2.4.2 Rasterizing geochemical data

The geochemical data are rasterized by pull-down menu "Rasterize point data" in the ENVI software. The output projection was determined, and output X/Y size was selected as 1000 m, meaning the spatial resolution of the formed rasterized images is 1000 m. Linear interpolation (quintic polynomial) was chosen. Smooth quintic

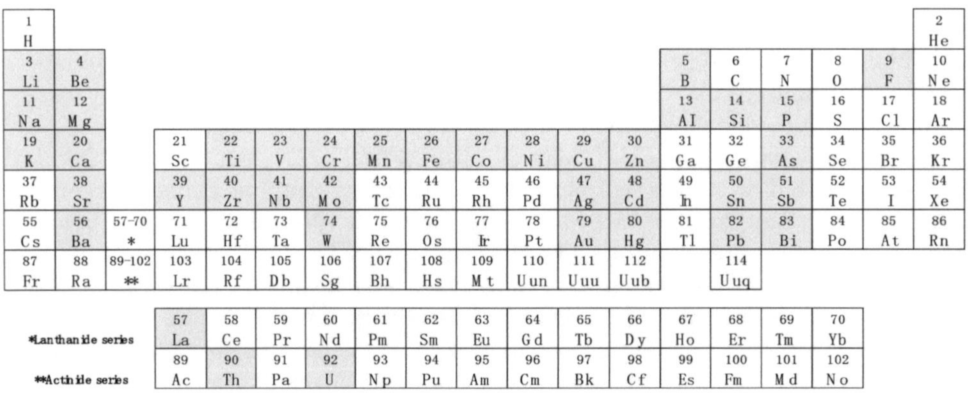

Figure 2.
Position of the elements measured by 1:2,000,000 regional geochemical survey in the periodic table, with shaded area as measured elements.

polynomial interpolation is performed by giving binary interpolation of Z values and smooth surface fitting at points that are irregularly distributed on the X-Y plane. The interpolation function is a fifth-degree polynomial in X and Y in a triangular cell and each polynomial is determined by the given values of Z and estimated values of partial derivatives at the vertexes of the triangle [32]. After the above steps, the image of a single element can be formed. And in the same way, the images of 39 elements can be created.

The sampling sites were irregular and the rasterized image covered a whole area in a rectangle in the process of rasterizing geochemical data (**Figure 3**). These inappropriate image contents can be eliminated by using the method of masking. To mask the incorrect area, the buffer zones of sampling site were used.

The formation of the buffer zone is that the geochemical sampling sites were overlaid by the ROI (region of interest) sites. The overlaid 5231 sampling sites were shown in **Figure 4**. By contrast, the maximum distance assigned to 2 pixels is much better to generate buffer zone. Buffer zone image was used to create an image mask. Finally the incorrect area was masked to generate the geochemical content image (**Figure 5**).

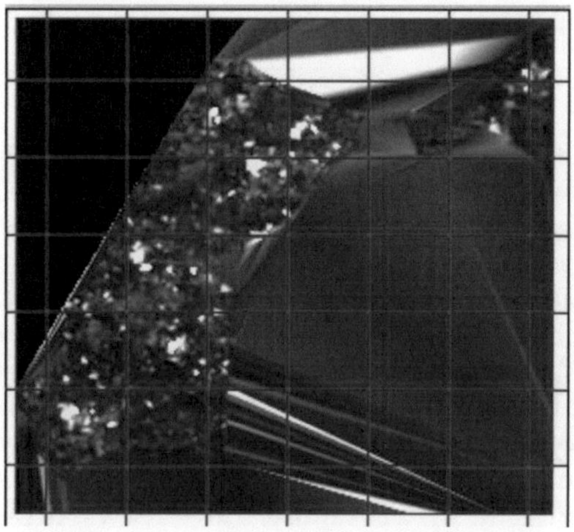

Figure 3.
Original Ag element grid map of Manzhouli area cut from northern Daxinganling metallogenic belt. Pixel size is 1000 m.

Figure 4.
Locations of sampling points in Manzhouli area. Sampling points were expressed in ROI and superposed on image maps as pixels. Pixel size is 1000 m.

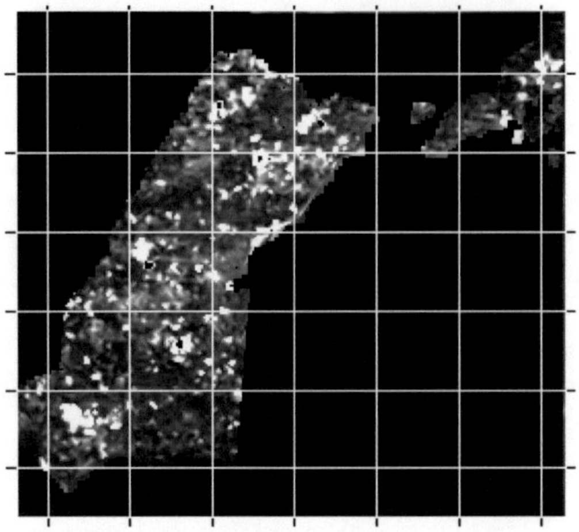

Figure 5.
Rasterized image of Ag element after a buffer zone mask.

3. Results

3.1 Building geochemical atlas

After 39 kinds of geochemical elements (or oxide) were generated, they would be put together to form an image atlas. The method is simple, namely, using "Laystacking" command, respectively, each image was successively overlayered together.

From the view of spectroscopy, the geochemical elements need to be classified. In the periodic table of the elements, elements of the same family possess similar chemical properties, and they have similar enrichment characteristics in the earth. In accordance with the periodic table, the element family was arranged from left to right. In each family, the order was arranged from top to bottom. In this way, the order of the arranged geochemical elements was as follows: Li, Na_2O, K_2O, Be, MgO, CaO, Sr, Ba, Y, La, Th, U, Ti, Zr, V, Nb, Cr, Mo, W, Mn, Fe_2O_3, Co, Ni, Cu, Ag, Au, Zn, Cd, Hg, B, Al_2O_3, SiO_2, Sn, Pb, P, As, Sb, Bi, and F.

3.2 Geochemical spectrum

In ENVI software, it is very easy to form the spectra which are constituted of the results of different geochemical elements. This paper defined these spectra as geochemical spectra, which is somewhat similar to the geochemical anomaly and the geochemical chart mentioned in geochemistry, all of which imply the content of geochemical element. All the data in the element content image are with original value, which is easy for data comparisons. If only considering the characteristic of the spectrum, methods of normalization may be adopted, namely, histogram stretching was conducted on each element content image to form the numerical range from 0 to 1, thus creating a clearer and more obvious contrast geochemical spectrum. **Figure 6** shows a comparison of the spectrum of main ore deposits in the Manzhouli region. The ore deposits shown in **Figure 6** are Sanhe lead-zinc deposit, Xiahulin lead-zinc deposit, Waixinhe molybdenum deposit, Babayi copper deposit, Wunugetushan copper-molybdenum deposit, Jiawula lead-zinc deposit, Chaganbulagen lead-zinc-silver deposit, and Erentaolegai silver deposit, respectively.

3.3 Image display and image enhancement

The rasterized geochemical image (**Figure 7**) may be carried out by image enhancement. For example, if expanding or changing the value field range of gray scale, or changing the distributional pattern of gray value, the sharpness of image may be enhanced. Some methods, e.g., direct gray transformation, histogram equalization, etc. may be adopted. And in order to make the edge of the image bright and clear, the image filtering method could be used. The image formed from geochemical data can constitute the ternary RGB image, e.g., the formed K_2O-Na_2O-SiO_2 image (**Figure 8**); it is known that K-Na-Si ingredient can be used to judge the composition of rocks.

3.4 Image statistics

Geochemical image can carry out a numerical statistics, which are somewhat different from the statistics of data of geochemical sampling sites. It is statistics of

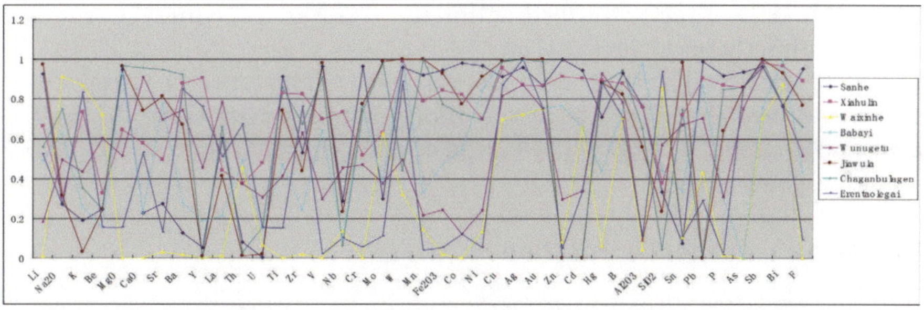

Figure 6.
Geochemical spectrum of typical deposits in Manzhouli region after histogram stretching.

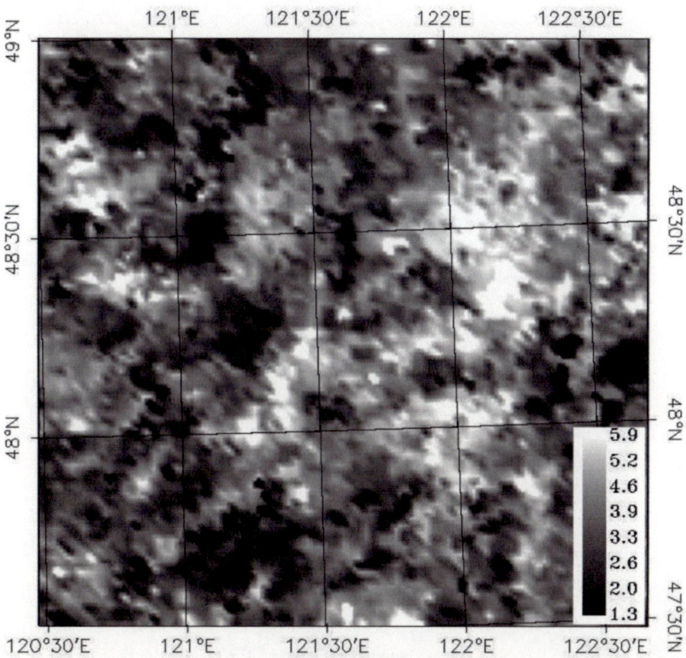

Figure 7.
Rasterized grayscale map of Na_2O element content in the middle segment of Daxinganling metallogenic belt.

all the pixels within the image. Basic statistics of a geochemical image involves the mean value, median, mode, range, contrast, etc.

Histogram is one of the important statistics of a geochemical image. Histogram refers to a discrete graph of probability density function of all gray values in the image, or it may be seen as a graphic expression of basic statistics of gray image. **Figure 9** is based on histogram and the chiefly related statistics. Under ENVI software, the calculation results of cumulative frequency can be obtained, and classification based on histogram analysis will be introduced in the next step.

Density slices to a gray geochemical image can create element anomalies. Cumulative frequency percentage can be used to determine anomalies or anomalies grading (**Figure 10**).

3.5 Algebraic operations and logic operations of image

Algebraic operations of image indicate that the corresponding image pixels of two (or more than two) of input images received four arithmetic operations, which in order are addition, subtraction, multiplication, and division. The algebraic operation cannot be directly fulfilled within the vector maps, while the rasterized maps can be directly performed.

Logical operations of images are widely applied, for instance, the masking method mentioned above used logic operations to form a mask band. A specific value in a pixel could be obtained by logical operations, and then a simple classification could be generated.

3.6 Geochemical image classification

In the vector image, undoubtedly, the anomaly image of elements is one of the final products in geochemistry. The anomaly map of elements may give users vivid

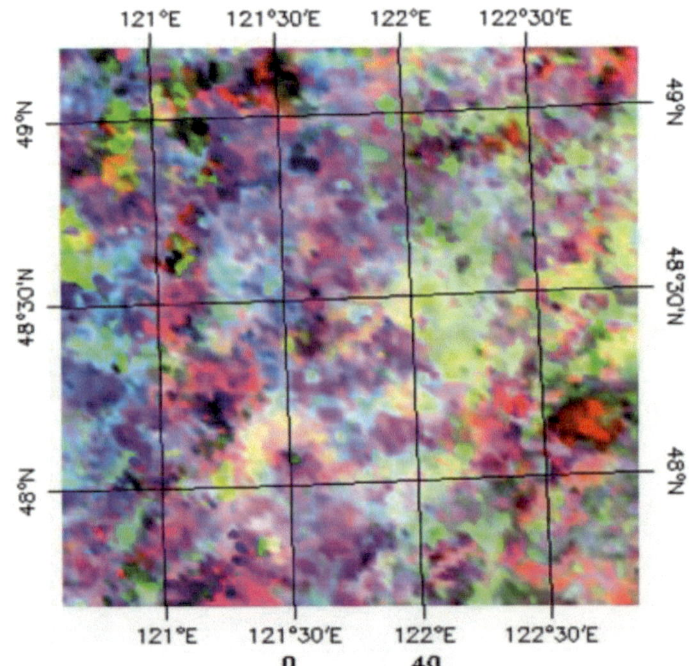

Figure 8.
K_2O, Na_2O, and SiO_2 ternary color image synthesis in the middle segment of Daxinganling metallogenic belt.

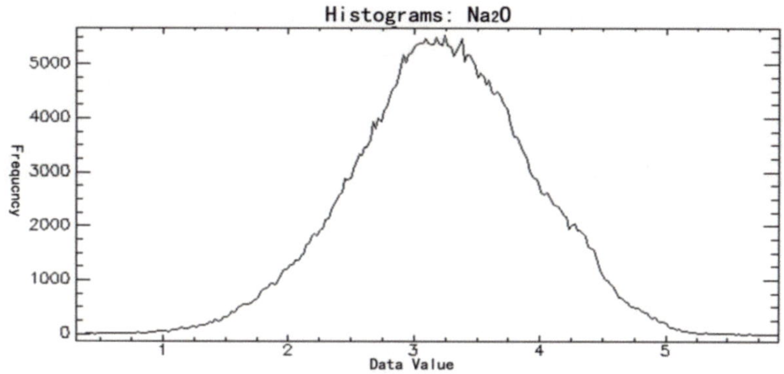

Figure 9.
Numerical statistical histogram of Na_2O content in the middle segment of the Daxinganling metallogenic belt.

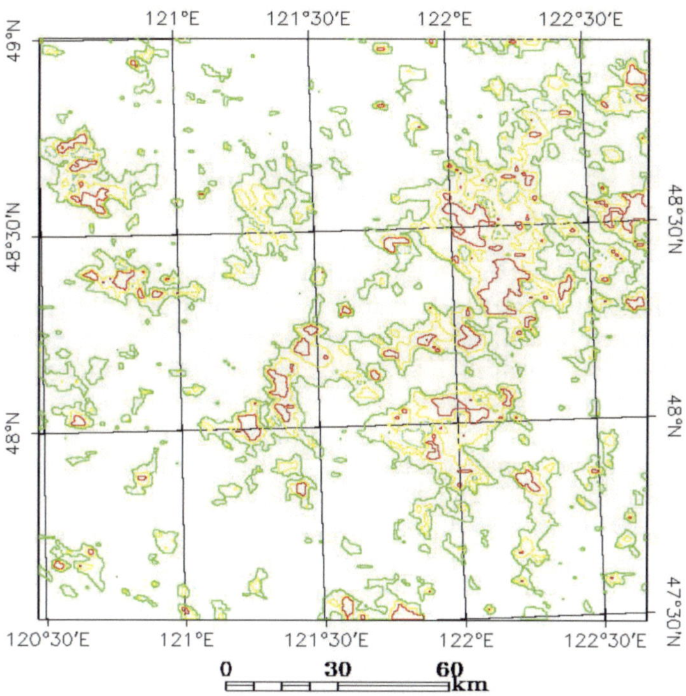

Figure 10.
Anomaly map of copper element formed in density slice in the middle of the Daxinganling metallogenic belt. According to histogram cumulative frequency statistics, anomalies were graded to 75% (green line), 85% (yellow line), and 95% (red line).

visual impression. Thus prospecting researchers can directly use the geochemical anomaly maps to explore the interested target. The results expressed in the rasterized image can also be fully used so as to employ the statistical method of density slicing. **Figure 10** is a density sliced map which was created by histogram statistics of copper element, and its result is similar to the geochemical anomaly map. Their difference is that the final rasterized image was irregularly dentate if enlarging a small area.

What is mentioned above is the simplest classification in the rasterized geochemical image, and they were based on the sole element anomaly. Most of the time, the classification using remote sensing images is divided by supervised one and unsupervised one, and their difference is that the supervised classification firstly gives category, whereas the unsupervised one is determined by the statistics characteristics of image data itself. The classification method used for remote images are suitable for the geochemical atlas. Usually employed methods include multilevel slice classifier, decision tree classifier, minimum distance classifier, maximum likelihood classifier, and the like (e.g., method of fuzzy theory, expert system method, etc.). SAM method mentioned later is one of the supervised classification methods.

4. Application of geochemical image in geology

The formed geochemical atlas can provide the prospecting target area just like conventional geochemical method and may also conduct multielement geochemical analysis. The geochemical image can accomplish the structural interpretation, e.g., linear structure and ring structure in geology just like what is fulfilled by the optical remote sensing. This paper does not restate these traditional methods but will mainly introduce the following three kinds of application in geology in the northern Daxinganling metallogenic belt.

4.1 Assisting in geological mapping

The geochemical atlas of 39 geochemical elements was generated in the northern Daxinganling metallogenic belt, including major elements and trace elements. The full use of all the elements will better assist geological mapping. Especially, in the Daxinganling Mountains, the outcrop is scarce because of the forest cover and that the field work of geological mapping encounters a great deal of difficulties. Therefore, boundaries of the geological bodies are indistinct, and the final boundaries are somehow judged by subjective experience. To employ unsupervised classification method may provide the reference for determining the boundaries of rock in the working area. As shown in **Figure 11**, the Chaihe area in the northern Daxinganling metallogenic belt was taken as an example; this working area belongs to stream sediment survey of the 1:200,000 Wuchagou sheet. The 39 geochemical element images are classified by K-Mean classification, and the geological interpretation map is created as the following one.

It can be seen that the geochemical mapping (**Figure 12**) may relatively clearly distinguish $\gamma_5^{2(2)}$ alkali feldspar granite from monzogranite. However, the boundary is different from that in the geological map (**Figure 11**). In the north and south, it was verified; but in the east of the map sheet, the rock which was delineated by geochemical images (**Figure 12**) was not presented in the geological map (**Figure 11**). Other Wuchagou basalt can also be easy to identify; two signs were manifested in the north, same as the geological map. Because Baoshi formation and Fujiawazi formation are volcanic, it is sometimes difficult to classify them. As a result, the interpenetration phenomenon is frequent. In the field work, it is hard to distinguish the volcanic rocks. For example, both Fujiawazi formation and Baoshi formation contain tuff; sometimes, the difference between intermediate lava and acidic lava is weak in the field. In this case, the divided geological map is worse than the geochemical classification.

4.2 Prospecting target selection

There is plenty of research on the methods of the prospecting target selecting using data-driven and knowledge-driven modes. In the past, selecting prospecting area was primarily based on the anomaly of the major ore-forming elements. The area with high anomaly value of a single element or integrated anomalies was selected

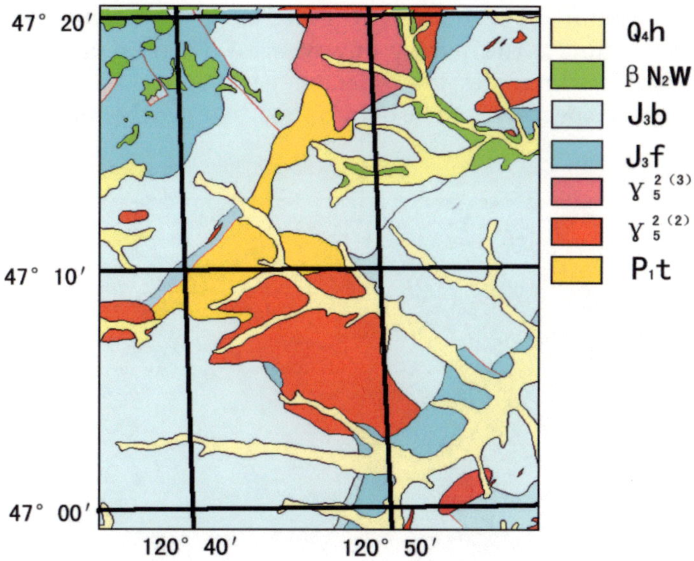

Figure 11.
Geological sketch of Chaihe area (1:200,000).

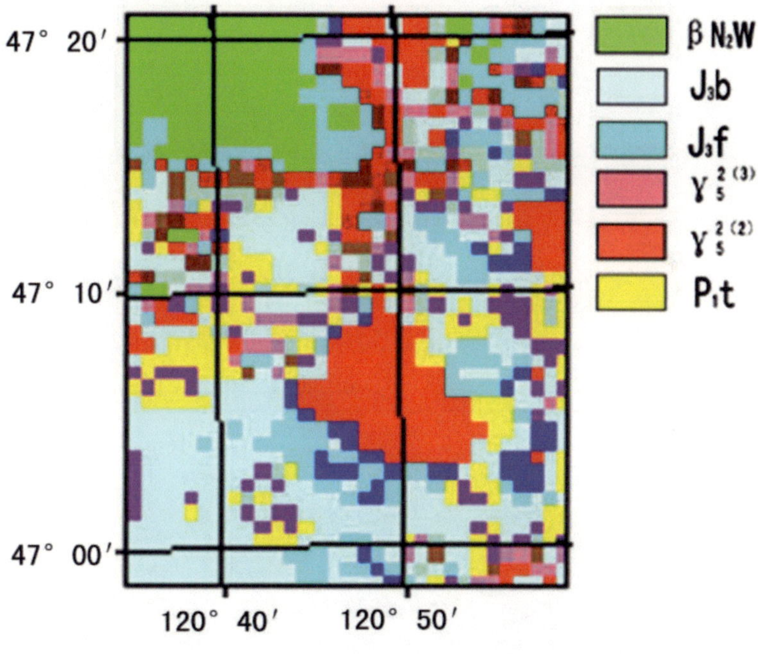

Figure 12.
Map of each unit of K-Mean classification of 39 geochemical elements in Chaihe area (legend codes are the same as **Figure 12***).*

as prospecting target. Although the large area of geochemical working had been carried out, fewer researchers utilize all elements for prospecting target selecting.

Combining the characteristics of the geochemical atlas in the northern Daxinganling metallogenic belt, the geochemical spectrum method may be adopted to exhibit the similarity with the known deposits on target locating. The most frequently used method is spectral angle mapper method (SAM) [33, 34]. SAM method utilize N-dimensions angle to match image elements and reference spectra. The geochemical spectra were regarded as vector, whose number of dimensions is the same as the number of waveband. Then using the angle algorithm for calculating the angles' inter-element geochemical spectra, the similarity of two geochemical spectra could be determined. The geochemical spectra of

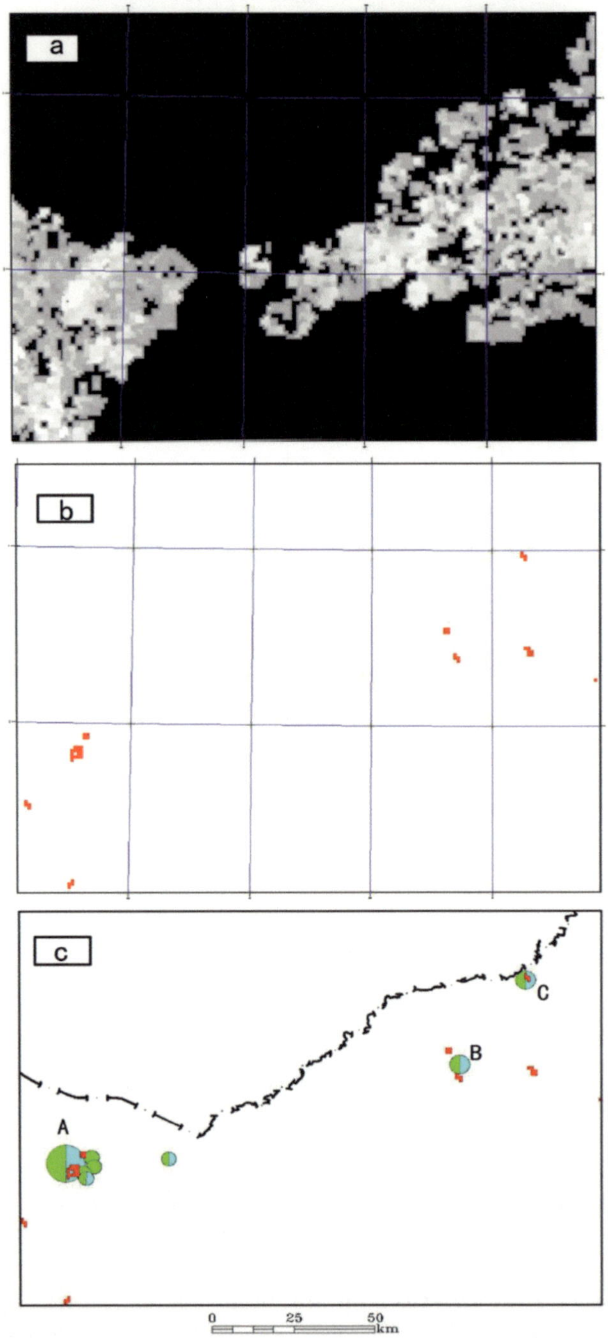

Figure 13.
Comparison map between the prospecting target and actual deposits in the Manzhouli region. The prospecting target was obtained by applying spectra angle method to some porphyry copper-molybdenum deposit; a is spectral angle map; b is the prospecting target formed by threshold segmentation of spectral angle map; and c is the corresponding location map between prospecting target and actual deposit. A is Wunugetushan deposit, B is Babayi deposit, and C is Badaguan deposit.

locations on the known deposits are regarded as end-member spectra, and then SAM is used to compare end-member spectra with the angles of each pixel vector in N-dimensions space. The smaller angle indicates that it fits better with the geochemical spectra of the discovered deposits. This method fully utilized the information of geochemical spectra and makes every elements involved in the classification. Additionally it emphasizes the shape characteristics of the geochemical content and greatly reduces the information such as the main ore-forming elements.

In this study, the geochemical spectral of the Wunugetushan copper deposit was taken as reference spectra, the SAM method was adopted, and the classification results have been verified by deposits of the same type (**Figure 13**).

4.3 The classification on mixed rocks

Since the late 1950s, Webb and his colleagues presented to collect fine granular sedimentary from drainages which stands for the average content of the catchment basins [35]. The subsequent regional geochemistry survey mainly based on their theory and method, namely, the sample collected, may stand for the contribution of all matters in the surrounding area of this sampling position. This is the same as to the so-called mixed spectra in remote sensing. Because large areal distributional mixed pixels evidently affect the calculation and classification of the remote sensing image, many researchers put forward the method of decomposing mixed pixels. Nowadays, methods of decomposition of mixed pixels are mainly classified into two classes, one is the linear spectral decomposition, which is based on the linear additivity of brightness of pixels, and the other is the fuzzy decomposition method.

In the process of geological mapping, the stratigraphic unit needs to be divided, and it includes various kinds of rocks. The Manitu formation on the Xiaodonggou section in the northern Daxinganling metallogenic belt served as an example. The standard strata, which were distributed between upper Baiyingaolao formation and lower Manketouebo formation, are 690.6 m thick. From bottom to top, the section involves green andesite (101.6 m), light gray andesitic-rhyolitic breccia tuff (219 m), dark gray, yellow gray andesite interlayered with debris tuff (190.3 m), and purple gray-dark andesite (179.7 m). In the fieldwork, it is difficult to observe all the rock types mentioned due to a few outcrops. As a result, the stratigraphic division can only be based on the limited artificial outcrops. Under this condition, the method of decomposing mixed pixels was used. Through decomposing the mixed pixels, the shares of various kinds of rocks can be achieved; thus it can assist stratigraphic unit classification in geological mapping.

5. Discuss and future prospects

In the past, regional geochemistry has made significant achievements in geology and mineral exploration. However, all of these relied on vector data, and the number of geochemical elements is limited, which narrowed the application of geochemical data. This paper only aims to supplement and modify the shortcomings of previous methods, rather than to overthrow or criticize the achievements attained by them.

The rasterized geochemical image possesses many advantages. The geochemical image is vivid for the visual interpretation. Additionally, data can be compatible for statistical analysis. That vectorized geochemical data accomplished can be achieved by the rasterized data in most cases. Furthermore, the imaged geochemical data could be processed with hyperspectral tools, which cannot be used in vector data.

The shortcoming of rasterized geochemical images mainly lies in that the raster format occupy a relative larger data storage space, and if the sampling sites is sparse, and the spatial resolution is set largely, a clear lattice shape will be displayed.

The increased geochemical density makes this kind of method to get more in-depth application. No doubt geochemical survey with larger scale can provide more information. China recently carries out geological survey on main metallogenic belts, and their sampling density was bigger. The sampling density in the northern Daxinganling metallogenic belt was averagely 4–8 sites per km^2 in the scale of 1:50,000; therefore the sampling density has been greatly increased. Followed by reducing the analysis data of geochemical elements, the usual analyzed elements now are Au, Ag, Cu, Pb, Zn, As, Sb, Hg, W, Sn, Bi, Mo, and so on; the purpose is for mineral exploration. With the increase of sampling data in unit area, the spatial resolutions of geochemical image will increase. The following job is to merge 39 geochemical elements of 1:200,000 with geochemical elements in 1:50,000 to create the multielement geochemical atlas with a relative higher resolution.

Integration with other types of geoscience data is also imperative. The geological map can finally transform to a rasterized image. The strata, magmatism, and so forth may be assigned values through various kinds of logical operations in rasterized image. Regional geophysical survey, for instance, aeromagnetic, airborne gravity, geomagnetic, gravity, and regional electrical method, may form the corresponding rasterized image. These data combined with the geochemical data will undoubtedly increase the information content of geosciences; therefore, it will develop a broader approach for intensive geological study and the comprehensive application of geosciences data.

6. Conclusions

Regional geochemical data of 1:200,000 in the northern Daxinganling area were rasterized using a method that triangulates a planar set of points. Consequently, a multilayered image database containing 39 elements/oxides was formed. The images were enhanced using an image enhancement technique and algebraic operations. The images were handled as multidimensional vector data. Accordingly, hyperspectral tools could be used for the processing. The geochemical signatures of deposits were extracted from the images. Enriched and depleted elements were distinguished by comparing them with regional geochemical statistics. The geochemical signatures represented the geochemical characteristics of ore deposits. The rock types were classified using the K-Means method, which assisted in the regional geological mapping, especially in the areas of dense forest. The geochemical signature of a typical ore deposit was processed by SAM, which determined the similarity between the deposit and pixels in the region. The prospecting target area was determined according to the angle. With increased geochemical data sampling density, as well as further integration with other geophysical, geological, and remote sensing data, rasterized geochemical images can be fully used in the future.

Acknowledgements

This research was supported by funds from the Chinese Ministry of Science and Technology (grant no. 2016YFC0600103) and the "Comprehensive integration and service of mineral geology and its metallogenic regularity in China" project of the China Geological Survey (grant no. DD20160346). In the process of drafting this paper, Prof. Liu Suhong, Prof. Huang Miaofen of Beijing Normal University, and other experts gave me a lot of help. We have benefited from discussions of relevant contents with researcher Zhang Yujun of the Chinese Academy of Geological Sciences. We would like to acknowledge the people mentioned above for their constructive and thoughtful comments that helped us to improve the paper.

Author details

Jiang Chen*, Zhaoxia Mao, Yan Zhao, Chunpeng Zhang and Shan Jiang
Shenyang Geological Center, CGS, Shenyang, Liaoning, China

*Address all correspondence to: chen07761@itc.nl

References

[1] Bogoch R, Shirav M, Beyth M, Halicz L. Geochemistry of ephemeral stream sediments in the Precambrian mountainous arid terrain of southern Israel. Journal of Geochemical Exploration. 1993;**46**(3):349-364

[2] Brantley SL, White AF. Approaches to modeling weathered regolith. Reviews in Mineralogy and Geochemistry. 2009;**70**(1):435-484

[3] Cheng Q, Bonham-Carter G, Wang W, Zhang S, Li W, Qinglin X. A spatially weighted principal component analysis for multi-element geochemical data for mapping locations of felsic intrusions in the Gejiu mineral district of Yunnan, China. Computers & Geosciences. 2011;**37**(5):662-669

[4] Hao LB, Lu JL, Li L, Mo GS, Yan GS, Shi YX, Zhao YY. Method of using regional geochemical data in geological mapping in shallow overburden areas. Geology in China. 2007;**4**:023 (in Chinese with English abstract)

[5] Wang W, Zhao J, Cheng Q, Liu J. Tectonic–geochemical exploration modeling for characterizing geo-anomalies in southeastern Yunnan district, China. Journal of Geochemical Exploration. 2012;**122**:71-80

[6] Zhao J, Wang W, Cheng Q. Application of geographically weighted regression to identify spatially non-stationary relationships between Fe mineralization and its controlling factors in eastern Tianshan, China. Ore Geology Reviews. 2014;**57**:628-638

[7] Drury SA, Drury SA. Image Interpretation in Geology. London: Blackwell science; 2001. pp. 1-290

[8] Gupta RP. Remote Sensing Geology. Springer Science & Business Media; 2013

[9] Bedini E. Mapping lithology of the Sarfartoq carbonatite complex, southern West Greenland, using HyMap imaging spectrometer data. Remote Sensing of Environment. 2009;**113**(6):1208-1219

[10] Berger BR, King TV, Morath LC, Phillips JD. Utility of high-altitude infrared spectral data in mineral exploration: Application to northern Patagonia mountains, Arizona. Economic Geology. 2003;**98**(5):1003-1018

[11] Bierwirth P, Huston D, Blewett R. Hyperspectral mapping of mineral assemblages associated with gold mineralization in the Central Pilbara, Western Australia. Economic Geology. 2002;**97**(4):819-826

[12] Kozak PK, Duke EF, Roselle GT. Mineral distribution in contact-metamorphosed siliceous dolomite at Ubehebe Peak, California, based on airborne imaging spectrometer data. American Mineralogist. 2004;**89**(5-6):701-713

[13] Rockwell BW, Hofstra AH. Identification of quartz and carbonate minerals across northern Nevada using ASTER thermal infrared emissivity data—Implications for geologic mapping and mineral resource investigations in well-studied and frontier areas. Geosphere. 2008;**4**(1):218-246

[14] Rowan LC, Mars JC. Lithologic mapping in the Mountain Pass, California area using advanced spaceborne thermal emission and reflection radiometer (ASTER) data. Remote Sensing of Environment. 2003;**84**(3):350-366

[15] van Ruitenbeek FJ, Cudahy T, Hale M, van der Meer FD. Tracing fluid pathways in fossil hydrothermal systems

with near-infrared spectroscopy. Geology. 2005;**33**(7):597-600

[16] Windeler DS, Lyon RJP. Discrimination dolomitization of marble in the Ludwig skarn near Yerington, Nevada using high-resolution airborne infrared imagery. Photogrammetric Engineering and Remote Sensing. 1991;**57**(9):1171-1177

[17] Chabrillat S, Pinet PC, Ceuleneer G, Johnson PE, Mustard JF. Ronda peridotite massif: Methodology for its geological mapping and lithological discrimination from airborne hyperspectral data. International Journal of Remote Sensing. 2000;**21**(12):2363-2388

[18] Harris JR, Rogge D, Hitchcock R, Ijewliw O, Wright D. Mapping lithology in Canada's Arctic: Application of hyperspectral data using the minimum noise fraction transformation and matched filtering. Canadian Journal of Earth Sciences. 2005;**42**(12):2173-2193

[19] Launeau P, Girardeau J, Sotin C, Tubia JM. Comparison between field measurements and airborne visible and infrared mapping spectrometry (AVIRIS and HyMap) of the Ronda peridotite massif (south-west Spain). International Journal of Remote Sensing. 2004;**25**(14):2773-2792

[20] Rivard B, Zhang J, Feng J, Sanchez-Azofeifa GA. Remote predictive lithologic mapping in the Abitibi Greenstone Belt, Canada, using airborne hyperspectral imagery. Canadian Journal of Remote Sensing. 2009;**35**(sup1):S95-S105

[21] Roy R, Launeau P, Carrere V, Pinet P, Ceuleneer G, Clenet H, et al. Geological mapping strategy using visible near-infrared–shortwave infrared hyperspectral remote sensing: Application to the Oman ophiolite (Sumail Massif). Geochemistry, Geophysics, Geosystems. 2009;**10**(2):1-23

[22] Van der Meer FD, Van der Werff HM, van Ruitenbeek FJ, Hecker CA, Bakker WH, Noomen MF, et al. Multi- and hyperspectral geologic remote sensing: A review. International Journal of Applied Earth Observation and Geoinformation. 2012;**14**(1):112-128

[23] Björklund A, Gustavsson N. Visualization of geochemical data on maps: New options. Journal of Geochemical Exploration. 1987;**29**(1):89-103

[24] Gustavsson N, Lampio E, Tarvainen T. Visualization of geochemical data on maps at the Geological Survey of Finland. Journal of Geochemical Exploration. 1997;**59**(3):197-207

[25] She HQ, Li HH, Li JW, Zhao SB, Tan G, Zhang DQ, et al. The metallogenetical characteristics and prospecting direction of the copper-lead-zinc polymetal deposits in the northern-central Daxing'anling Mountain, Inner Mongolia. Acta Geologica Sinica. 2009;**83**(10):1456-1472 (in Chinese with English abstract)

[26] Miao L, Zhang F, Fan WM, Liu D. Phanerozoic evolution of the Inner Mongolia–Daxinganling orogenic belt in North China: Constraints from geochronology of ophiolites and associated formations. Geological Society, London, Special Publications. 2007;**280**(1):223-237

[27] Zhang X, Zhang H, Tang Y, Wilde SA, Hu Z. Geochemistry of Permian bimodal volcanic rocks from central Inner Mongolia, North China: Implication for tectonic setting and Phanerozoic continental growth in Central Asian Orogenic Belt. Chemical Geology. 2008;**249**(3):262-281

[28] Zhang L, Gao B, Li W, Chen Z, Sakyi PA, Jin X. Early Mesozoic

tectono-magmatic activity and mineralization in Northeast China: Evidence from Re–Os to U–Pb studies of the Taipingchuan porphyry Cu–Mo deposit in the Derbugan metallogenic belt. International Geology Review. 2014;**56**(15):1837-1851

[29] Xuejing X, Huanzhen S, Tianxiang R. Regional geochemistry-national reconnaissance project in China. Journal of Geochemical Exploration. 1989;**33**(1):1-9

[30] Cai YZ, Feng WH. A review of 1:200 000 regional geochemical exploration in the forest-swamp area of eastern inner Mongolia. Geophysical & Geochemical Exploration. 2003;**27**(6):423-424

[31] Wang HF, Ye ZC. The application of new 1:200 000 regional geochemical methods to Wuchagou sheet in Inner Mongolian forest-swamp area. Geophysical & Geochemical Exploration. 2003;**27**(6):435-437

[32] Akima H. A method of bivariate interpolation and smooth surface fitting for irregularly distributed data points. ACM Transactions on Mathematical Software (TOMS). 1978;**4**(2):148-159

[33] Hecker C, Van der Meijde M, van der Werff H, Van der Meer FD. Assessing the influence of reference spectra on synthetic SAM classification results. IEEE Transactions on Geoscience and Remote Sensing. 2008;**46**(12):4162-4172

[34] Kruse FA, Lefkoff AB, Boardman JW, Heidebrecht KB, Shapiro AT, Barloon PJ, et al. The spectral image processing system (SIPS)—Interactive visualization and analysis of imaging spectrometer data. Remote Sensing of Environment. 1993;**44**(2):145-163

[35] Webb JS. Regional geochemical reconnaissance in the Namwala Concession area, Zambia (No. 47). 1964

Chapter 5

Geochemical Methods to Assess Agriculture Sustainability

Trolard Fabienne, Kaniewski David and Bourrié Guilhem

Abstract

Facing global changes and the challenge of food security, scientists are being questioned by decision-makers and stakeholders on the sustainability of agro-systems. The main difficulty in dealing with this question is to obtain enough data over long periods of time. Monitoring slow drifts and weak noises is needed to forecast tipping points that can jeopardize the present steady state. High-resolution datations by radiocarbon coupled with detailed palynological determinations in sediments, historical archives on yields and crop quality, and high-frequency field in situ measurements give information on climatic changes from multi-secular to seasonal and hourly time scales. In the long term, climatic forcing dominates agriculture performance, at that time only organic agriculture, with oscillations between prosperity and misery driven by climate and intermediated by civilization flourishing and collapsing; in the medium term, in modern agriculture, irrigation provides a provisional buffering effect on yield and crop quality despite present warming; in the short term, either under non-fertilized forested ecosystem or intensive rice cropping, the same patterns are evidenced and point to the importance of soil microflora shifting from aerobiosis to anaerobiosis. In all cases, geochemistry offers appropriate tools to decipher the climate-soil-agriculture complex interplay.

Keywords: geochemistry, sustainability, agro-systems, isotopes, in situ monitoring

1. Introduction

In the world, agriculture developed in the last 10,000 years, and most plant and animal species of interest to humans have been domesticated since. Today's agriculture still relies almost exclusively on these same species, and a large part of economy is based upon the production and trade of only less than 10% of these domesticated plant and animal species [1].

Facing global changes and the challenge of food security, scientists are being questioned by decision-makers and stakeholders in the territories on the sustainability of agro-systems. Valuable information on this topic and recommendations can be derived from the study of practices and processes related to agriculture over time.

All farming systems implemented since 10,000 years on the Earth are still present today. But is it reasonable to consider that the most efficient systems in terms of productivity can be the most sustainable? How can we approach and explore the notion of sustainability in agriculture? In current agricultural practices, is it possible to distinguish between those that could be considered sustainable and those that would not be? Sustainability means maintenance of steady-state conditions over long periods. The main difficulty in dealing with these questions is to obtain enough data over long periods of time. The ability to detect slow drifts and weak noises is needed to forecast tipping points that can eventually modify or jeopardize the present steady state.

In the present chapter, geochemical data are processed to gain this information. In fact, the interest of geochemistry is to produce scalar-type information that obeys the universal laws of chemistry, whatever the place and time in the Earth's surface conditions. It offers remarkably robust concepts and models that are valid for several orders of magnitude on space and/or time scales (e.g., from nanometer to megameter and from picosecond to millennium). Thus, we have used it in the study of different agro-systems and exploring long-time (up to 1000 years), medium-time (around 60 years), and short-time (1 hour to 3–4 months) phenomena.

2. Short historical evolution of agriculture

Agriculture appeared almost concomitantly in the Middle East, particularly in the Euphrates Valley, in Central America, and in Asia. With the global warming ending the last quaternary ice age, agriculture spread to Europe, West Africa and South America, and then North America and Oceania. From a historical perspective, we can observe (**Figure 1**), firstly, that the development of agrarian systems is directly linked to the growth of human population and, secondly, that each positive inflection of the curve corresponds to significant improvements and innovations in agricultural practices.

During the evolution of agricultural practices, we distinguish a first period from less than −10,000 years to 1850, called organic agriculture, in which progresses rested mainly on the improvement of the use of internal energy of the agriculture system. It thus starts with manual farming where only the farmer's labor force counts. Then the lightly hitched agriculture appears in which the animal force is introduced but with animals not specifically dedicated to work in the fields (soil tillage) and, finally, the heavily hitched agriculture, with animals selected for their physical strength, in which the farmers could maintain them during winter, because they became able to stock a stable surplus of fodder grown during the summer season. In the middle of the nineteenth century, with organic farming, a farmer could work a maximum of 5 ha with a yield remaining below 50 quintals per hectare.

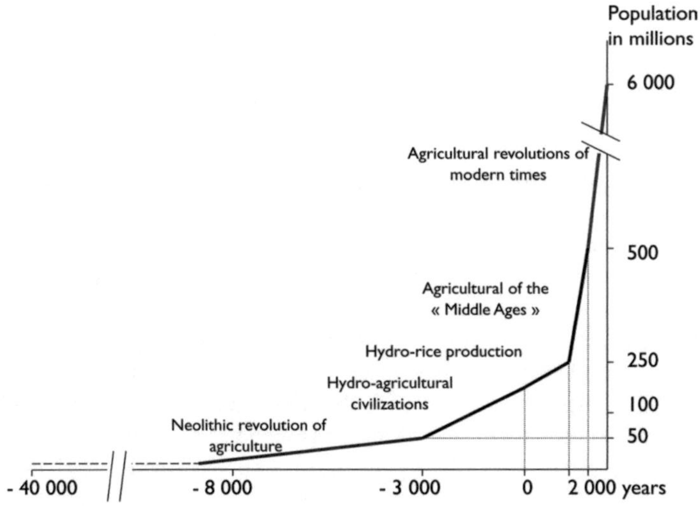

Figure 1.
The development of agrarian systems is linked to the growth of the human population (from [1]).

The second period, called mineral agriculture, runs from 1860 to today. It is marked by the increased use of energy external to the agriculture system, with the steam engine for the heavily hitched agriculture of the end of the nineteenth century and especially with oil that allowed the development of mechanization, irrigation, and production of fertilizers and xenobiotics and with genetic improvement. Thus, for the most efficient farming systems in the world, a farmer is able to work more than 150 ha and to produce over 20,000 quintals per year. In the present world, however, all types of agriculture still coexist.

3. Investigations of the period from 4500 years BP to today

To illustrate the long-time investigations, the period from 4500 years to today has been explored by the help of radiocarbon dating and palynological determinations.

3.1 Materials and methods

For isotopic and palynological characterizations, soil and/or sediment cores were drilled and sampled as continuously as possible. One core was drilled on the delta of Mirna River (Gulf of Venice) in coastal Croatia continuously at 720 cm depth [2].

The chronology of the core is based on accelerator mass spectrometry ^{14}C dating of short-lived terrestrial samples (seeds and small leaves). No botanical macro-remains were found in the middle core, but the use of marine shells (involving the radiocarbon-dating reservoir effect) and bulk samples (involving potential contaminants) was strictly avoided in order to minimize chronological biases in the age-depth model. Dated samples were calibrated (1 sigma (σ) and 2σ calibrations, respectively, 68 and 95% confidence interval) using CALIB REV 7.0.4 [3]. The average chronological resolution for the core stratigraphy is 7 years per cm^{-1} (1.43 mm per $year^{-1}$). **Figure 2** shows some examples of micro-botanical remains like pollen and macro-botanical remains like seeds, kernels, and wood pieces.

3.2 Results and discussion

Since the rise of agriculturally based societies in the Mediterranean, and concomitant population growth in coastal areas, humans have gradually generated irreversible impacts on natural biotic resources. The introduction of agricultural practices and human-induced fires in northern Istria is dated to 5000 BP [4]. The cultivated species were mostly cereals (*Secale*, *Hordeum*, *Avena*, and *Triticum*), the same genus found in the Mirna region (**Figure 3**). The succession of agropastoral activities can be determined here, with cereals (about 3000 years BP), olive growing, viticulture, and orchards (about 2000 years BP) (**Figure 3**). Optima appear in Medieval and Roman eras.

The periodicity of agropastoral activities was investigated using a wavelet analysis, highlighting the long-term trends versus storm surges with a 950-year period [2]. This suggests that low storm activity and enhanced freshwater inputs in the delta have favored arboriculture and agriculture. Conversely, periods of higher storm surges, which generated the intrusion of saline water into the freshwater-fed plains and into the groundwater table, led to severe agricultural losses. The comparison of the two signals, fitted to a 950-year filter, shows that, at a millennial time scale, anthropogenic activities and storminess are in antiphase [2]. In addition, we can observe that the most prosperous periods of agropastoral activities correspond to the optima of the Roman civilization and the Medieval era. The abandonment of

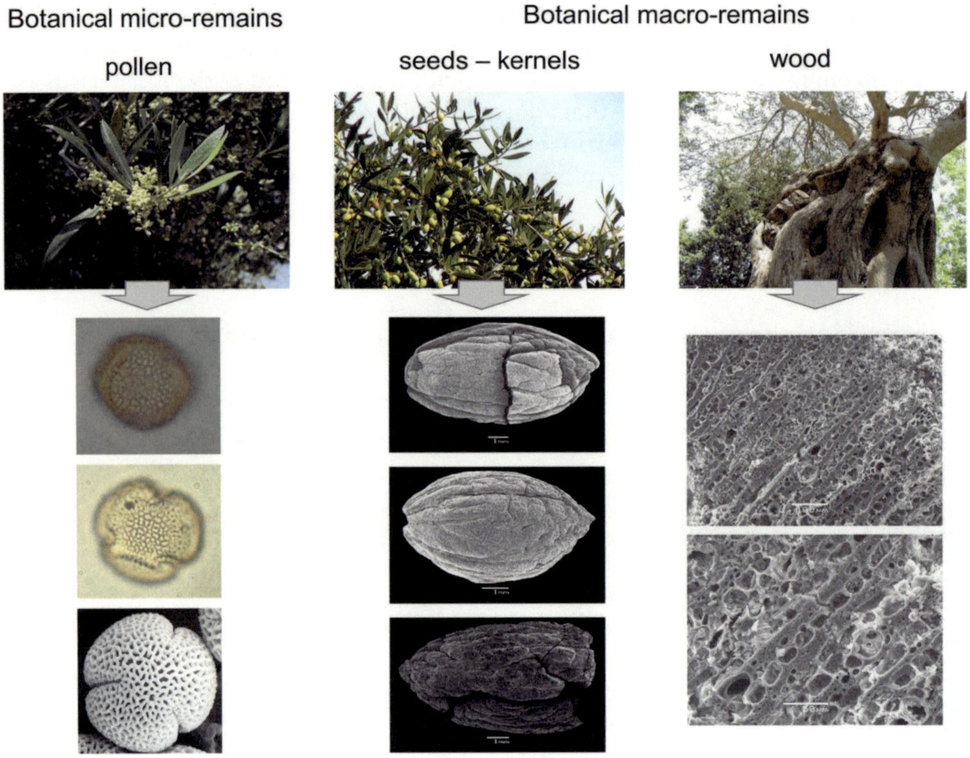

Figure 2.
Examples of botanical micro- and macro-remains observed by optical microscopy or by scanning electronic microscopy.

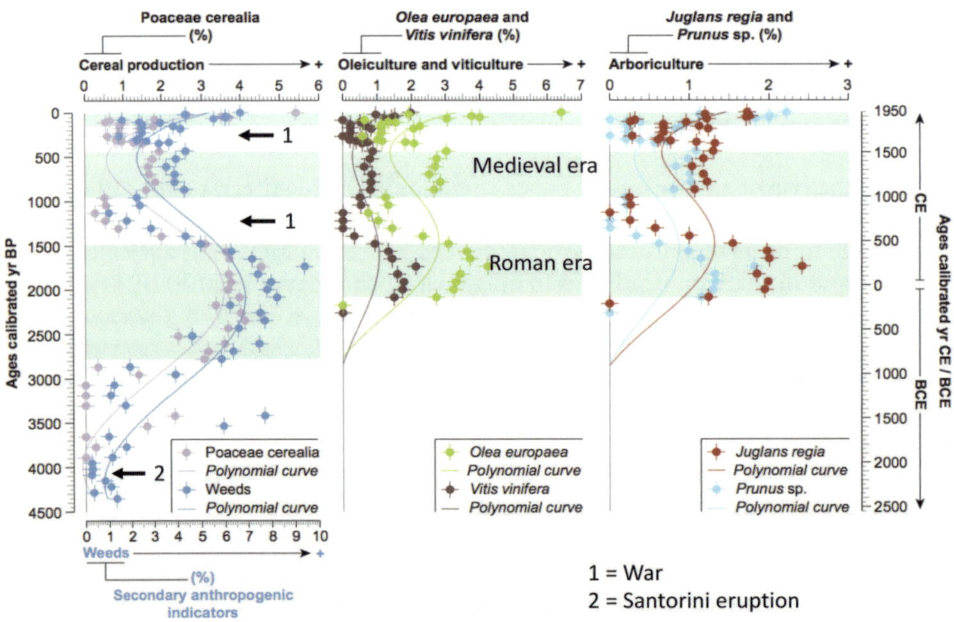

Figure 3.
Reconstruction of agropastoral activities from Mirna region during 4500 years with the help of isotopic ^{14}C datation and palynological determinations (modified from [2]).

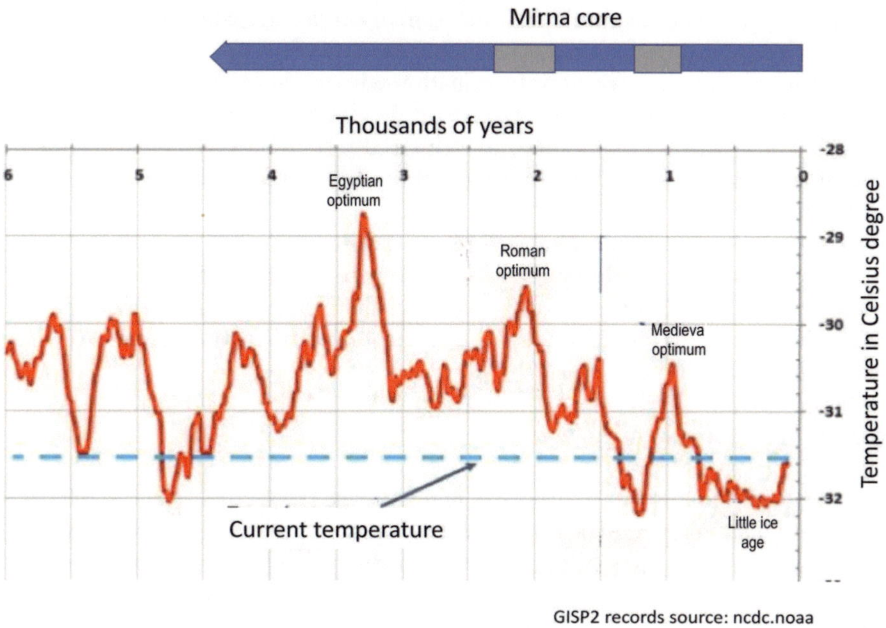

Figure 4.
Position of intensive paleo-agropastoral activities (in gray) from Mirna core placed on the 6000 years BP temperature variation reconstructed from ice cores drilled in Central Greenland. For the detail of datation of Mirna's core, see ***Figure 3****.*

all agricultural activities around 1650 years BC can be ascribed to the consequences of the major volcanic eruption of Santorini. The ends of the prosperous periods can be related to the invasions of the northern peoples (the so-called Dark Ages) and war periods. A maximum is indeed observed at 3500 BP, followed by a minimum at 3200 BP. This latter minimum can be ascribed to the collapse of Bronze Age civilization near 1177 BC = 3127 BP [5].

By comparison with the curve of temperature over 6000 years BP building up from ice cores drilled in the center of Greenland, the results show that the agropastoral optima of Mirna's core correspond also to thermal optima (**Figure 4**).

4. Investigations of medium-time period (60 years) in intensive agriculture

For the medium-time investigations, the consequences of intensive agriculture of the last 60 years have been studied in the irrigated grasslands in Crau's area (hay production with COP label) in southeast of France.

4.1 Materials and methods

The Crau's area covers up to 600 km^2 at the south of Alpilles mountains. A part of the territory is occupied by a natural semiarid steppe, named "coussoul," and, in another part, by irrigated grasslands and orchards. Thanks to Adam de Craponne, a sixteenth-century engineer, a first irrigation canal was built up, and the irrigation network extended until the nineteenth century. The network supports the production of the Crau hay with a protected designation of origin, which is exported all over the world to feed racehorses, the Sisteron lamb, and the Arles Merinos sheep, which are produced with a label of geographical indication of origin. Irrigated grassland production is regulated by three cuts of hay (i.e., cut 1, cut 2, and cut 3) and sheep grazing in the field during the winter. The gravity irrigation with water of Durance river secures the renewal, up to 70%, of the groundwaters in the aquifer, which supply the water consumption of 280,000 inhabitants and industrial activities of Marseille harbor.

A database concerning hay's mineral content, dry matter, and climate dynamics was statistically analyzed [6].

In addition, the geochemical variations of water composition during its pathway from irrigation channel till the phreatic aquifer were recorded and modeled. For modeling with PHREEQC software [7], we have defined a priori the processes, which can impact water chemistry, such as evaporation, dissolution, precipitation, and exchange with the plant [8].

4.2 Results and discussion

Nitrogen and phosphorus contents in hay increase from cut 1 to cut 3, whereas the potassium decreases significantly in cut 3. These results are explained by seasonal changes in the floristic composition of the hay [9]. But globally the total inorganic content in Crau hay increases over time from cut 1 to cut 3, and this order has remained constant since 1960 (**Figure 5**).

Statistically, these results show a steady state of the production, which has been maintained, both in quantity and quality despite an average temperature increase of 1.9°C since 1960 [10].

For each chemical element measured in the waters, a model of fluxes is built up [6]. Activities and saturation indexes ($SI = \log Q - \log K$) were computed by using PHREEQC [7], using phreeqc.dat database: activity coefficients were computed with Debye-Hückel extended law, as ionic strength is small enough (*ca.* 0.01 M). The reaction of reduction of nitrate into ammonium was removed from the database as it is biologically mediated, and N(III) and N(V) were considered as distinct elements separated by a kinetic barrier.

On **Figure 6a** models for calcium are presented. At each step, where chemistry of water changes, the soil solution is computed. The time step is 2–3 months corresponding to the duration of hay growth for one cut. Thus, between solution S1 and S2, the evaporation of the water induced a loss of water and a concentration of the elements. Then, from S2 to S3, the pCO_2 (partial pressure of CO_2) of the soil, which is 30–100 times larger than in the outer atmosphere, results in acidification of solution and calcite dissolution. From S3 to S4, the model

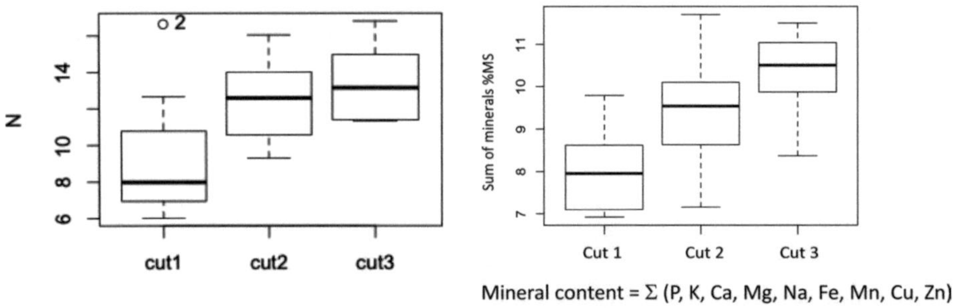

Figure 5.
Nitrogen and total inorganic element contents in hay as function of the cut (from [9]). The mineral content of the hay is defined, per unit of dry matter expressed in percentage, by the sum of the content of the following elements: phosphorus, potassium, calcium, magnesium, sodium, iron, manganese, copper, and zinc.

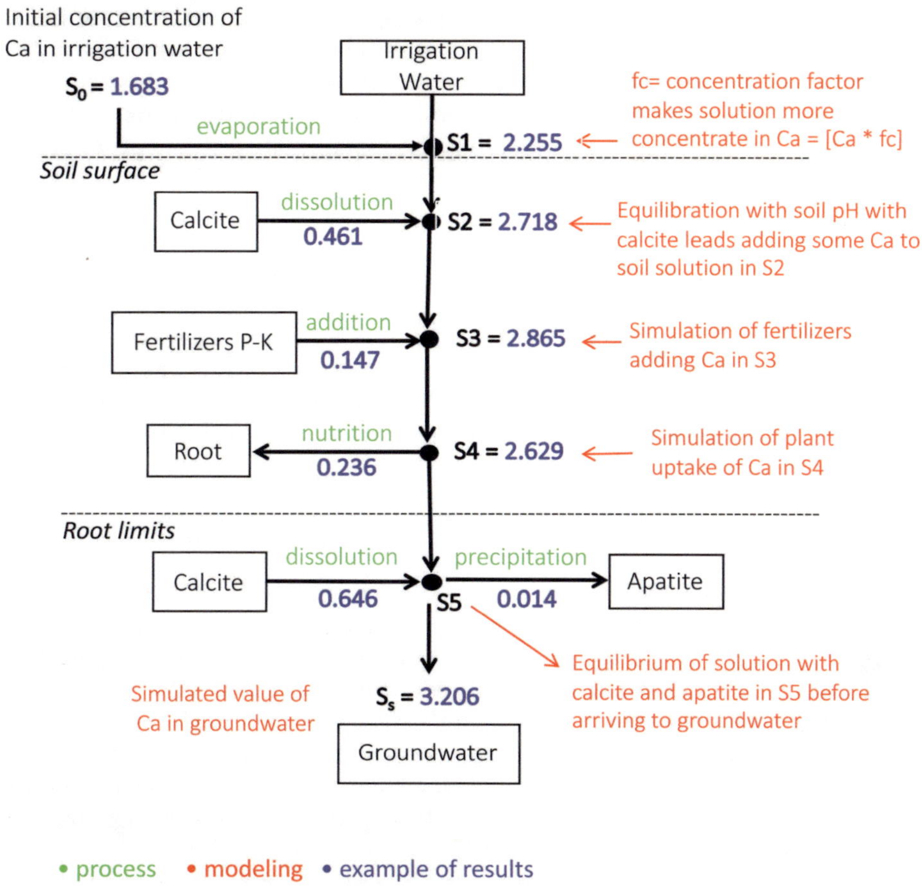

Figure 6.
(a) The pathway of calculation of fluxes of Ca during soil solution changes from surface irrigation water to groundwater; (b) for the 4 years of monitoring of water in Crau's area, comparison between measured and computed values of Na, Ca, Mg, C, K, S, and Cl for each cut (from [6]).

simulated the fertilizer impact on the soil solution. Inorganic fertilizers (P, K) consist of gypsum $CaSO_4.2H_2O$, calcium dihydrogenphosphate $Ca(H_2PO_4).2H_2O$, arcanite K_2SO_4, and sylvite KCl. The last three minerals were introduced in the database, with their thermodynamic properties [7C]. Dissolution of fertilizers was simulated by PHREEQC as the dissolution of a mixture of the above minerals. From S4 to S5, the model simulated the element uptake by plants. P absorption by plants was simulated as the removal of calcium phosphate from the solution, S absorption by plants as the removal of gypsum, and calcium being absorbed in excess to the sum of P and S, the remaining Ca absorption was simulated as a CaO removal from the solution; Na, K, and Mg absorption by plants were simulated respectively as the removal of Na_2O, K_2O, and MgO from the solution. Removal of elements by plant is computed by PHREEQC as a dissolution with negative coefficients, in the same way as evaporation is computed with a negative coefficient for water. To avoid transient negative concentrations, fertilizer dissolution was simulated before absorption by plants. In point S5, the soil solution is reequilibrated with the minerals of the aquifer.

The fluxes have been computed for the three cuts per year during 4 years of monitoring the water quality both in irrigation network and in groundwater (**Figure 6b**). All simulations are computed at the average temperature of groundwater. With a R^2 larger than 0.96, regardless of the year or the cut, or the chemical element, the proposed model simulates very well the transformation of irrigation water into groundwater, describing for these 4 years a steady state.

Thus, our findings suggest that irrigation, both with the water inputs and quality of water, have played a key role for the sustainability of hay production till 60 years.

5. Investigations of short time in agro- and ecosystems

The short time (1 hour intervals) of processes in agro- and ecosystems were recorded by in situ monitoring both of solid and water in a forested hydromorphic soil and in a paddy field. In both systems, the geochemistry of iron is marked by strong interactions between the solid minerals and the soil solution when oxidoreduction phenomena occurred [11, 12].

5.1 Materials and methods

With the progress of in situ instruments and sensors developed for spatial research and Mars exploration, a miniaturized Mössbauer spectrometer, Mimos II, was built up [13, 14]. The signal emitted by the source of the spectrometer is recorded using reflection-based geometry, which is not influenced by the thickness of the sample, unlike conventional transmission geometry [15]. Mimos II was placed in a tube, in the soil. Mössbauer spectra were recorded in a nondestructive way through windows at different depths till 1.20 m depth. Periodically the instrument is moved back to fixed positions that allow us to monitor the changes of iron mineralogy as a function of time [15].

The progress of in situ instrumentation also concerns the monitoring of water quality thanks to progress in the oceanographic research. Such a probe was used to monitor soil water in Brittany in the Fougères forest and in Camargue paddy field. The pH, redox potential, temperature, and electrical conductivity were hourly measured, and data were stored in the probe and collected every fortnight.

5.2 Results and discussion

Figure 7a shows a typical Mössbauer spectrum obtained in the forested area, characterizing the fougerite mineral, which is a mixed hydroxide of Fe(II) and Fe(III) from green rust family [16, 17]. The points are the measures recorded by the instrument, and the lines are obtained by fitting Lorentzian functions to the signal. Thus the spectrum shows two doublets D1 and D2 characterizing the crystal environment of Fe(II) and one doublet D3 characterizing the crystal environment of Fe(III) in the mineral structure.

About 30 spectra were obtained during one hydrological year (i.e., between October 1998 and September 1999), and the variations of the ratio of Fe(III) to total iron are reported as a function of depth (**Figure 7b**). At a given depth, it changes in time, with fluctuations of the water table and anaerobiosis/aerobiosis conditions. When in reductive conditions, the signal of fougerite appears in less than 1 week [15].

Both at Fougères and in Camargue, plots of Eh-pH diagrams (**Figure 8**) show fast and large variations in short time in a quasi-identical range of variations, though soils are largely different, i.e., acidic in Brittany and carbonated in

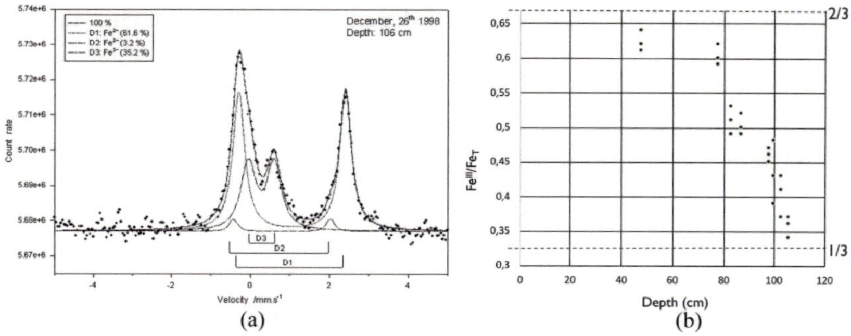

Figure 7.
(a) Example of Mössbauer spectrum obtained in situ with Mimos II instrument; (b) variation of the Fe^{III}/Fe_{total} *ratio of iron minerals as function of depth and time (from [15]).*

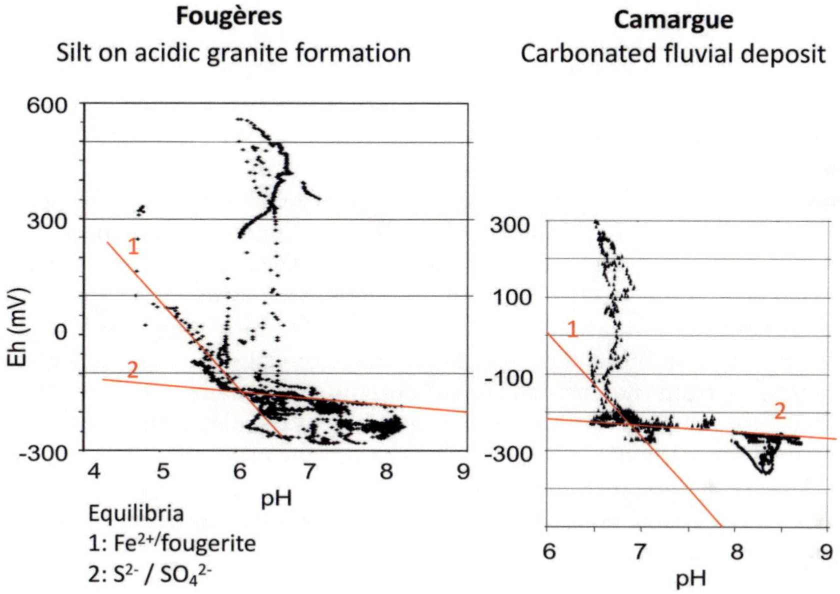

Figure 8.
Eh-pH variations in soil solution at Fougères and in Camargue recorded in situ with an oceanic probe. Each point is the average of sequence of 20 measurements recorded during 1 mn every hour.

Camargue. The observed measures both at Fougères and in Camargue show large variations covering the whole domain of existence of aqueous Fe(II) and Fe(III) oxides. The rH values calculated according to the following equation:

$$rH = Eh/0.029 + 2\,pH \tag{1}$$

vary in between 4 and 14.4 at Fougères and in between 4 and 5.5 during the irrigation period in the Camargue. These values are much smaller than 20, which is the upper limit admitted for reducing media [18]. The "loops" of fast evolution, in a few hours, toward oxidizing conditions (i.e., Eh up to 100 mV) with small variations of pH are explained by the infiltration of oxygen-rich rainwater and then a return to the redox state of the initial steady state.

The representation of the thermodynamic equilibria of Fe^{2+}/fougerite and of S^{2-}/SO_4^{2-} is given by curve 1 and curve 2, respectively. The geochemical modeling shows that the principal controls of the water quality are exerted by Fe^{2+} versus green rust fougerite, when reduction is moderate, and sulfide versus sulfate equilibria, when reduction is high.

6. Conclusions

Agriculture evolving from organic to mineral farming since 10,000 years has greatly increased its performance and has capacity to feed humanity. Geochemistry combined with other disciplines, such as palynology, bioclimatology, mineralogy, etc., offers cutting edge methods to investigate both long-time, medium-time, and short-time processes in agro- and ecosystems.

The long-term investigations show that soils and sediments have recorded agropastoral activities and thus paleo-agriculture can be studied. The combination of global climatic events (temperature deduced from ice cores), storm activities, and relative sea level movements led to socioeconomic destabilization of the local society and impacted the sustainability of agriculture in these remote times. This has led to a renewal of historical studies, integrating the above cited disciplines and others, especially genomics with DNA analyses and identification of mutations, anthropology, and archeology. Emblematic events in the history of mankind were recently revisited: late Bronze Age collapse of civilization mentioned above [5], shedding light on the mysterious invasions of the "Peoples of the Sea," and very recently the fall of the Roman Empire [19]: here too, e.g., isotope geochemistry (^{10}Be) was used to reconstruct variations of solar irradiance. According to this study [19], climatic changes combined with apparition of three successive pandemias (first pest due to *Yersinia*, under emperor Justinian) are facilitated by commercial routes connecting tropical regions in Africa to the Mediterranean, from the Caspian Sea to Scotland. Eventually, economic (including monetary), political troubles, and migrations of peoples from Asian steppes to the West due to climatic change led to wars and the ultimate collapse of the Roman Empire. It appears then that multidisciplinary studies nowadays extensively use geochemistry not only to precise datation but to reconstruct climatic changes and migration routes.

In the present concern about climate change, it is important to stress that it is multifactorial and such studies necessitate decades of cooperation. They point too to the factors that increase the resilience of societies toward external forcings and help to detect tipping points and to build scenarios that can be integrated in decision support systems [20].

Medium-term investigations, particularly for agro-systems, are difficult to perform because data are scarce: the time step for an agronomist or a farmer is most often the cultural year. The medium-term investigations of the grasslands in Crau area show that Craponne's irrigation network has built up the soils in Crau by silt depositions since the sixteenth century. Since 60 years, the hay quality remains constant. Water quality remains constant along the year and contributes directly to the plant nutrition. This irrigated agro-system has mitigated until today the temperature increase, around 2°C, observed in this area.

In situ instrumentation and sensors allow us to address dynamics at two scales that escape to classical agronomic analysis: slow drifts and high-frequency events (pulses). In particular the continuous monitoring of master variables of soil solution can record these events and help us to understand the dynamics of agro-systems and their evolution in the long term.

This kind of information is needed to manage the sustainability of agro-systems because the soil is a nonrenewable capital at human scale.

Acknowledgements

The authors thank the following PhD students for their contribution: L. Cary, F. Feder, and G. Mohammed. They thank also M. Mayor, A. Régeard, and G. Sévenier for their help as technical support for chemical analysis and field monitoring.

Author details

Trolard Fabienne[1*], Kaniewski David[2,3] and Bourrié Guilhem[4]

1 INRA-AU, UMR 1114 Emmah, Avignon, France

2 IUF, Paris, France

3 Paul Sabatier University, ECOLAB, Toulouse, France

4 Académie d'Agriculture de France, Paris, France

*Address all correspondence to: fabienne.trolard@inra.fr

References

[1] Mazoyer M, Roudart L. Histoire des Agricultures du Monde. Du Néolithique à la Crise Contemporaine, Seuil; 2002. 736 p. EAN 9782020530613

[2] Kaniewski D, Marriner N, Morhange C, Faivre S, Otto T, Van Campo E. Solar pacing of storm surges, coastal flooding and agricultural losses in the Central Mediterranean. Scientific Reports. 2016;**6**:25197. DOI: 10.1038/srep25197

[3] CALIB 14C Calibration Program. 2018. Available from: http://calib.qub.ac.uk/calib/ [Accessed: December 20, 2018]

[4] Kaniewski D, Marriner N, Morhange C, Rius D, Carre MB, Faivre S, et al. Croatia's mid-late Holocene (5200-3200 BP) coastal vegetation shaped by human societies. Quaternary Science Reviews. 2018;**200**:334-350. DOI: 10.1016/j.quascirev.2018.10.004

[5] Cline E. 1177 BC: The Year Civilization Collapsed. New Jersey, USA: Princeton University Press; 2014. ISBN 978-0-691-14089-6

[6] Mohammed G, Trolard F, Gillon M, Cognard-Planck A, Chanzy A, Bourrié G. Combination of a crop model and a geochemical model as a new approach to evaluate the sustainability of an intensive agriculture system. Science of the Total Environment. 2017;**595**:119-131. DOI: 10.1016/j.scitotenv.2017.03.146

[7] Parkhurst DL, Appelo CAJ. User's guide to PHREEQC (version 2)—A computer program for speciation, batch-reaction, one-dimensional transport, and inverse geochemical calculations. In: Water-Resources Investigation Report. Denver: U.S. Department of the Interior, U.S. Geological Survey; 1999. pp. 99-4259

[8] Bourrié G, Trolard F, Chanzy A, Ruget F, Lecerf R, Charron F. Sustainable intensive agriculture: Evidence from aqueous geochemistry. Procedia Earth and Planetary Science. 2013;**7**:93-96. DOI: 10.1016/j.proeps.202.03.077

[9] Mohammed G, Trolard F, Bourrié G, Gillon M, Tronc D, Charron F. A long-term data sequence (1960-2013) to analyze the sustainability of hay quality in irrigated permanent grasslands under climate change. American Journal of Agriculture and Forestry. 2016;**4**(6):140-151. DOI: 10.11648/j.ajaf.20160406.11

[10] Olioso A, Lecerf R, Bailleux A, Chanzy A, Ruget F, Banton O, et al. Modelling of drainage and hay production over the Crau aquifer for analyzing impact of global change on aquifer recharge. Procedia Environmental Sciences. 2013;**19**:691-700. DOI: 10.1016/j.proenv.2013.06.078

[11] Stumm W, Sulzberger B. The cycling of iron in natural environments: Consideration based on laboratory studies of heterogeneous redox processes. Geochimica et Cosmochimica Acta. 1992;**56**(8):3233-3257. DOI: 10.1016/0016-7037(92)90301-X

[12] Trolard F, Bourrié G. Chapter 5: Geochemistry of green rusts and fougerite: A reevaluation of Fe cycle in soils. Advances in Agronomy. 2008;**99**:227-287. DOI: 10.1016/S0065-2113(08)00405-7

[13] Klingelhöfer E, Morris RV, Bernhardt B, Rodionov D, de Souza PA, Squyres SW, et al. Athena MIMOS II Mössbauer spectrometer investigation. Journal of Geophysical Research, Planets. 2003;**108**(E12):8067. DOI: 10.1029/2003JE002138

[14] Rodionov D, Klingelhöfer G, Bernhardt B, Schröder C, Blumers M, Kane S, et al. Automated Mössbauer spectroscopy in the field and

monitoring of fougerite. Hyperfine Interactions. 2006;**167**(1-3):869-873. DOI: 10.1007/s10751-006-9371-9

[15] Feder F, Trolard F, Klingelhöfer G, Bourrié G. In situ Mössbauer spectroscopy—Evidence for green rust (fougerite) in a gleysol and its mineralogical transformations with time and depth. Geochimica et Cosmochimica Acta. 2005;**69**:4463-4483. DOI: 10.1016/j.gca.2005.03.042

[16] Trolard F, Abdelmoula M, Bourrié G, Humbert B, Génin JMR. Mise en Évidence d'un Constituant de type "Rouilles Vertes" dans les sols Hydromorphes—Proposition de l'Existence d'un Nouveau Minéral: La "Fougérite", Séries IIa. Vol. 323. Paris: Comptes Rendus de l'Académie des Sciences; 1996. pp. 1015-1022

[17] Trolard F, Bourrié G, Abdelmoula M, Ph R, Feder F. Fougerite, a new mineral of the pyroaurite—Iowaite group: Description and crystal structure. Clays and Clay Minerals. 2007;**3**:323-334. DOI: 10.1346/CCM.2007.0550308

[18] IUSS Working Group WRB. Word reference base for soil resources. International soil classification system for naming and creating legends for soil maps. In: World Soil Resources Reports 106. Roma: FAO; 2014

[19] Harper, K. The fate of Rome, Climate, Disease, and the end of an Empire. Princeton, New Jersey, USA: Princeton University Press; 2017. 440 pp. ISBN: 9780691192062

[20] Trolard F, Reynders S, Dangeard M-L, Bourrié G, Descamps B, de Mordant de Massiac J-C. Territoires, Villes et Campagnes Face à l'Étalement Urbain et au Changement Climatique—Une Démarche Intégrative Pour Préserver les sols, L'eau et la Production Agricole. Paris: Editions Johanet; 2013. p. 143. ISBN: 979-1091089098

Chapter 6

Geoenvironmental Characterization of Sulfide Mine Tailings

Tomás Martín-Crespo, David Gómez-Ortiz and Silvia Martín-Velázquez

Abstract

Spain has a long mining tradition dating from pre-historic times up to the present day. The cessation of mining activity has generated a large amount of mine wastes, most of which represent geochemical hazards. Mine tailings are watery sludge composed of medium-to-fine-grained material, resulting from grinding and mineral processing (e.g., galena, pyrite, sphalerite, and arsenopyrite). They entail both an accumulation and a potential subsequent emission source of trace elements (i.e. As, Cu, Fe, Pb, and Zn) with formation of acid mine drainage (AMD). Mineralogical and geochemical techniques (in combination with geophysical surveys and aerial photographs studies) have been jointly applied to selected mine areas. Seven mine deposits from the most important mine districts in Spain have been selected: Iberian Pyrite Belt, Cartagena-La Unión, Alcudia Valley, and Mazarrón. The main goal is focused on getting a geoenvironmental characterization as complete as possible by determining the geometry, evolution in time and composition of mine ponds, and the possible occurrence of AMD, for identifying related environmental hazards.

Keywords: mine tailings, mineralogy, geochemistry, geoenvironmental, Spain

1. Introduction

Sulfide ore minerals are generally concentrated by milling and flotation, which produces tailings containing gangue minerals and residual sulfides. Milling involves crushing and grinding to reduce particle size and liberate ore minerals from the rock matrix. After ore minerals have been extracted and concentrated, the resulting tailings are commonly dewatered and deposited in sub-aerial tailings, impoundments, or stockpiles [1]. They are piled up as less than 5 cm thick layers and are slightly differentiated by compositional and/or granulometric features. Although the metal content is removed in the metallurgic process, some ore sulfides (e.g. pyrite, galena, sphalerite, chalcopyrite, arsenopyrite, etc.) can be deposited, either because they were not sufficiently high-grade for use, or due to a deficient extraction technology. They entail both an accumulation and a potential emission source of trace elements (e.g., Cu, Fe, Pb, and Zn). Oxidation of the sulfide minerals accumulated in the abandoned mine tailings may cause: (a) highly contaminating acid mine drainage (AMD) from leakages and (b) mobilization of significant quantities of trace elements such as As, Cd, Cu, Hg, and Pb. It becomes necessary to identify and characterize these hazardous areas where large quantities of potentially toxic elements can be released into the environment. Mine ponds are, therefore, an important environmental problem, especially if they are abandoned.

Spain has a long mining tradition dating from pre-historic times up to the present day. A large amount of mine installations, galleries, and waste deposits were abandoned until the 1980s by the cessation of mining activity. Pollution from these sources can originate via mining spills, leakages, or wind-blown dust, and toxic elements with a high mobility can cause huge environmental problems: accumulation in flora and fauna, reducing the quality of streams and groundwater.

An inventory of the abandoned mine waste deposits has been prepared by the Spanish Ministry of the Economy, through the Directorate General of Energy Policy and Mining [2]. The most relevant contribution of the inventory consists in the classification of the existing mine waste deposits based on their hazard potential for infrastructure and the human population. Another significant aim is focused on providing a qualitative geotechnical and environmental assessment of the elements at risk and an associated description. Further knowledge of the current status of the highest potential risk deposits is required because of the preliminary nature of the inventory, carried out by means of visual surveys and without sampling or testing.

In summary, a complete geoenvironmental characterization of the affected areas is crucial for any proposal of effective measures that could help to minimize environmental impact and concern.

2. Study zones

Seven mine areas from the most important metallic mining district in Spain have been selected (**Figure 1**): La Naya, Monte Romero and Mina Concepción from Iberian Pyrite Belt [3, 4], Brunita from Cartagena-La Unión [5], San Quintín from Alcudia Valley [6], and San Cristóbal and Las Moreras from Mazarrón [7].

2.1 Iberian Pyrite Belt

One of the largest concentrations of massive sulfide mineralizations in the world is hosting at the volcano-sedimentary rocks of the Iberian Pyrite Belt, in the southwest (SW) of the Iberian Peninsula. The main ore mineral is pyrite, although lower quantities of sphalerite, galena, chalcopyrite, and arsenopyrite are also found. An intense mining activity in this province is related to the exploitation of S, Cu, Pb, Zn, Ag, and Au from the sulfide ore minerals. Different studies have pointed out the significant concentrations of certain trace elements in sediments and soils surrounding mining or waste sites in the Iberian Pyrite Belt district [8]. The waters of the Tinto and Odiel fluvial systems are also affected [9]. Remediation has only been conducted in a few of the mine areas, although they still display significant environmental issues [4]. In 1998, an environmental disaster occurred in the SW of Spain, when the tailing dam of one of the bigger mines from the district was ruptured. Around 2×10^6 m^3 of heavy metal-bearing sludge and ~4×10^6 m^3 of acidic waters were released [10].

Monte Romero (**Figure 1**) comprises two mine ponds located at the Cueva de la Mora mine site, where Pb- and Zn-bearing minerals were benefited [3]. La Naya (**Figure 1**) is a mine pond located to the southeast of Minas de Riotinto town, and is one of the largest deposits mined during the extensive works in this mining group. The main ores extracted were Cu-bearing minerals and pyrite. Another mine

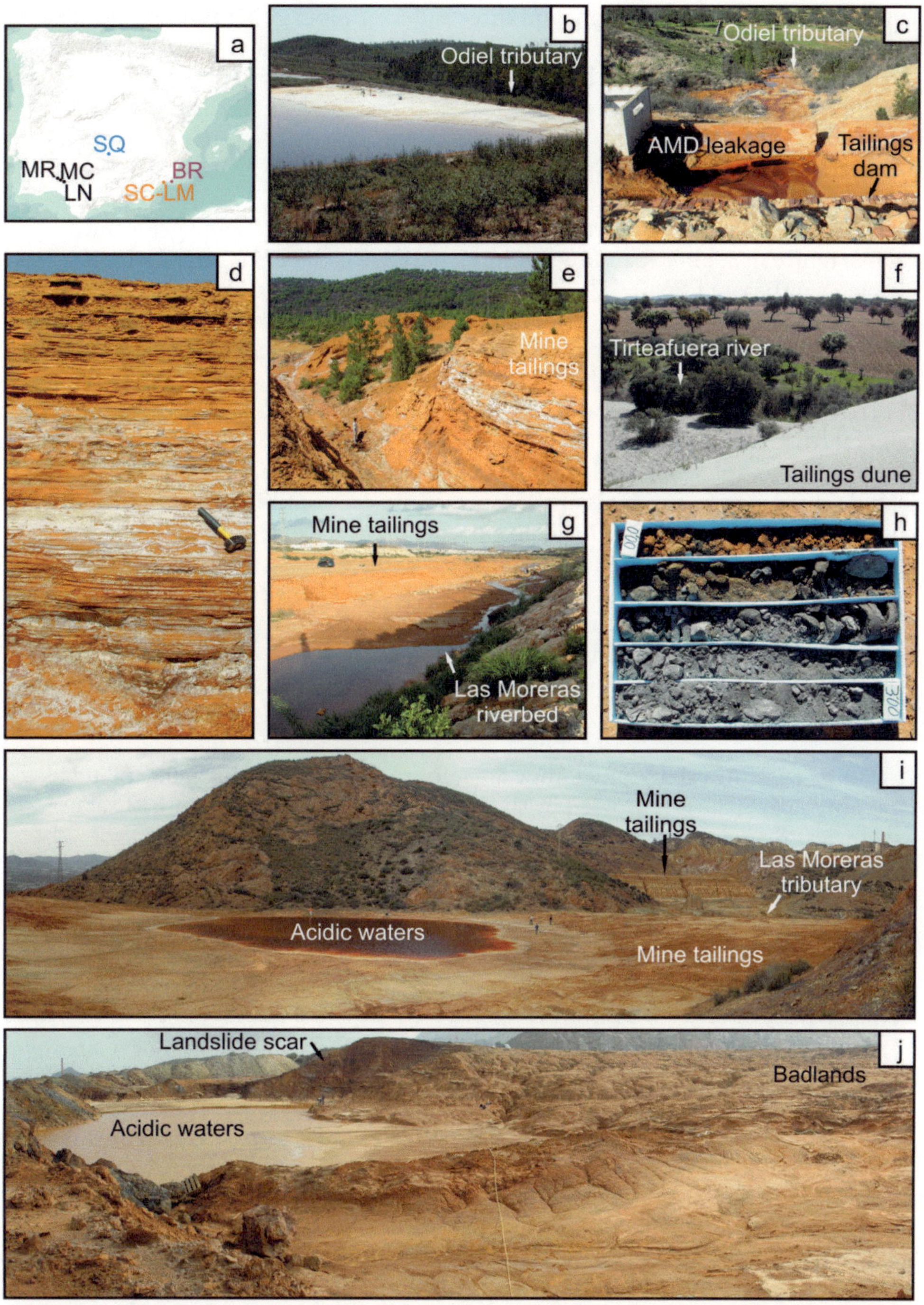

Figure 1.
Location and field photographs of the different studied mine areas: (a) location of the sulfide mine ponds in Iberian Pyrite Belt (LN—La Naya, MC—Mina Concepción, and MR—Monte Romero), Cartagena-La Unión (BR—Brunita), Alcudia Valley (SQ—San Quintín), and Mazarrón (SC—San Cristóbal and LM—Las Moreras); (b) Monte Romero; (c) acid leakage from Mina Concepción tailings; (d) detail of tailings in La Naya; (e) La Naya; (f) tailing dune in San Quintín; (g) Las Moreras; (h) borehole core samples from Brunita tailings; (i) San Cristóbal; and (j) Brunita.

site studied was Mina Concepción, a restored mine pond located on the SE of the Almonaster la Real village (**Figure 1**). Tailings from the metallurgical treatment and benefit of pyrite were piled up over metavolcanic lithologies [4]. The contention dyke retaining them is 8 m high, and three collector pipes going through the dyke controlled the drainage. The physical restoration and the landscape integration of the mine pond were achieved by sealing and reforestation with pine trees, substituting the original vegetation.

2.2 Cartagena-La Unión

One of the most significant places of geochemical pollution and geotechnical instability in Spain's abandoned mining heritage is the Cartagena-La Unión district, southeast of Spain [2]. The Brunita mine pond is one of the numerous tailing deposits in the district, affecting the surrounding watercourses that reach La Manga coastline, a major tourism location in SE Spain. Human beings, fauna, flora, groundwater, and agricultural soils are negatively stilted [5].

The mine tailings were produced from grinding and metallurgical treatment of mineral from Eloy and Brunite mines between 1952 and 1981 [2]. The main ore minerals are pyrite, sphalerite, galena, marcasite, and pyrrhotite. Other minor sulfides include arsenopyrite, minerals of the tetrahedrite-tennantite group, chalcopyrite, and stannite [11]. In October 1972, an extreme rainfall event caused damage in the Brunita mine pond (**Figure 1**). A tailing flash flood killed one person and caused serious material damage. As a result, a retaining wall of the coarse-grained tailings was built. Since 1981, when the mine was closed, no further works on restoration or reclamation have been carried out.

2.3 Alcudia Valley

The San Quintín abandoned mine area, located at the Alcudia Valley (Ciudad Real province, Spain), is crossed by the Don Quixote Route [12], a tourist set of itineraries created in 1995 to celebrate the IV Centenary of the publishing of "El ingenioso hidalgo Don Quijote de La Mancha". This route, the longest ecotourist route in Europe, was recently declared as Cultural Itinerary by the Council of Europe. It could soon reach the rank of Humanity Heritage because of its high cultural and environmental quality. These features make the San Quintín area a busy tourist route, and the environmental characterization of potential hazards is so necessary.

The ore mainly comprised Ag-bearing galena and sphalerite as major phases of a hydrothermal mineralization also including pyrite, chalcopyrite, marcasite, pyrrotine, bournonite, siderite, boulangerite, and ankerite [13]. The exploitation was performed from 1887 to 1934, date of the mining closure. In 1973, a new treatment plant was installed for re-working of approximately three million tons of tailings. Several tons of cinnabar from the Almadén mine (Ciudad Real, Spain) were experimentally treated in this new plant with successful results. At present, several mine tailings resulting from re-working together with the ruins of the mine structures are clearly visible. AMD from the tailings is recognized. Furthermore, a tailing dune, formed over one of the ponds, is migrating and toward the agricultural soils surrounding the mine area (**Figure 1**). The course of the Arroyo de la Mina stream, crossing the mining area, was altered and presently runs along the limits of the mine ponds.

2.4 Mazarrón

Mazarrón is located 4 km from the Mediterranean coast in SE Spain, and was one of the most important mining districts in the area [14]. It was exploited from Roman times to the early 1960s for Pb, Al, Ag, and Zn. Together with mining activities, the Mazarrón area is characterized by intensive farming and tourist pressure. Mining deposits caused significant water and soil pollution, and led to negative effects on both agricultural and tourism land uses. A correct geo-environmental characterization of the affected area is important for any proposal of restoration and remediation focused on minimizing environmental impacts.

The ponds are located on the hill slopes of the San Cristóbal and Los Perules hills, situated near a watercourse that drains to the Las Moreras watercourse, in turn flowing into the Mediterranean Sea (**Figure 1**). The main ore minerals were sphalerite, pyrite, and Ag-bearing galena. Other minor sulfides were arsenopyrite, chalcopyrite, the tetrahedrite-tennantite group, stibnite, cinnabar, and berthierite. The mine tailing ponds, near the Las Moreras dry watercourse, are situated on Quaternary alluvial and colluvial deposits (**Figure 1**). Although the total level amount of rainfall is not high, the area is subjected to strong stormy events each year, which can induce flash flooding phenomena.

3. Methodology

Mineralogical and geochemical techniques normally used to determine the composition of mine tailings, soils, waters, and watercourse sediments and the possible occurrence of AMD are described. In the case of high Hg contents, gaseous mercury emissions were analyzed too. Sampling features such as methods, sampling depth, analytical techniques, etc. are summarized in **Table 1** and described below.

3.1 Description of sampling

At Brunita, San Quintín, and San Cristóbal-Las Moreras areas, nondisturbed rock drill core tailing samples were collected from boreholes using a rotary drilling machine with a core bit diameter between 86 and 100 mm. Sampling was carried out by digging down below the surface of each pond, eliminating the surficial sealing to prevent falling material inside the borehole during drilling. Sampling depth of the unaltered samples varies between 0.5 and 1 m, depending on the borehole depth. All samples were air-dried for 7 days, passed through a 2-mm sieve, homogenized, and stored in plastic bags at room temperature prior to analyses. Below mine tailings, colluvial sediments (2–4 m) in San Quintín and watercourse sediments in San Cristóbal (2.5 m)—Las Moreras (4.5 m) were drilled, collected, and analyzed to obtain a complete geoenvironmental characterization of the area. Where a rotary drilling machine is not possible to place, samples were collected with an Eijkelkamp soil core manual sampler. Sampling was sequential with a centimeter vertical constant spacing and lower in depth than boreholes. This is the case of tailings studied at the Iberian Pyrite Belt district (**Table 1**). In many occasions, soils surrounding mine facilities show evident signals of contamination from different sources (tailings, ponds, open shafts, etc.) and pathways (wind erosion, water flows, etc.) affecting different receptors (agricultural soils, colluvial sediments, humans, etc.). In these scenarios, it is necessary to collect representative sample soils from the studied zone, and from a natural soil far enough of the mining area as a background sample (blank), in the case of San Quintín mine area. This is necessary to compare the potentially toxic element contents with the natural amount in the surrounding soils.

Water sampling is also necessary to check the metal amount in watercourse affected by the mining operations, as well as the possible AMD generation from the tailings. Several water samples have been collected depending on the features of the studied mine site: water sample collected at 8.5 m depth from a borehole

Mine district	Study mine	Ore mineralogy	Sample type	Sampling	Analytical techniques	Complementary techniques
Iberian Pyrite Belt	La Naya	**Py**, Sp, and Ga	Tailings	Manual sampler (2.7 m depth)	XRD, ESEM+EDX, INAA, pH	ERT (1 profile)
	Monterromero	**Sp**, **Ga**, and Py	Tailings	Manual sampler (2 m depth)	XRD, ESEM+EDX, INAA, pH	ERT (2 profiles)
	Mina Concepción	**Py**, Sp, Ga, Apy, Cpy, and Mg	Tailings	Manual sampler (1 m depth)	XRD, INAA	ERT (6 profiles)
			Water	Leakage	ICP-MS, pH, EC	
Cartagena-La Unión	Brunita	**Py**, **Sp**, **Ga**, Ma, Pyr, Te, Cpy, Apy, and St	Tailings	Borehole (24 m depth)	XRD, ICP-MS, pH, EC	ERT (6 profiles), aerial photographs
			Water	Water table	ICP-MS, pH, EC	
Alcudia Valley	San Quintín	**(Ag)-Ga**, **Sp**, Py, Ma, Cpy, and Pyr	Tailings/colluvial	Boreholes (10–12 m depth)	XRD, ICP-MS, ESEM+EDX, pH	ERT (7 profiles), aerial photographs
			Water	Borehole (8.5 m depth); watercourse; AMD	ICP-MS, pH	
			Soil	Surround; Blank		
			Air	Summer-winter	ZAAS-HFM	
Mazarrón	San Cristóbal	**Py**, **Sp**, **(Ag)-Ga,** Cpy, Apy, Te, Ci, and St	Tailings	Borehole (5.5 m depth)	XRD, ICP-MS, ESEM+EDX, pH	ERT (1 profile)
			Watercourse sediments			
			Water	Water table	ICP-MS, pH, EC	
	Las Moreras		Tailings	Borehole (6 m depth)	XRD, ICP-MS, ESEM+EDX, pH	
			Watercourse sediments			ERT (1 profile)
			Water	Watercourse	ICP-MS, pH, EC	

XRD, X-ray diffraction; ESEM, environmental scanning electron microscopy; EDS, energy dispersive X-ray; INAA, instrumental neutron activation analysis; ICP-MS, inductively coupled plasma-mass spectrometry; EC, electrical conductivity; and ZAAS-HFM, atomic absorption spectrometer with Zeeman effect.

Table 1.
Main sampling features of the studied mine district in Spain: Iberian Pyrite Belt, Alcudia Valley, Cartagena-La Unión, and Mazarrón [3–7].

(San Quintín), water samples from the tailing pond (Brunita and San Cristóbal), water sample from the watercourse crossing the mining area (San Quintín and Las Moreras), or water sample from the leakage of an abandoned (San Quintín) and a restored mine pond (Mina Concepción). All water samples were collected in 250 ml plastic bottles, were kept in a refrigerator at 4°C and, prior to analysis, and were filtered with 45 μm pore spacing. In San Quintín mine site (Alcudia Valley district), total gaseous mercury (TGM) was also measured using a 200 m sample spacing grid during both summer and winter by means of an atomic absorption spectrometer with Zeeman effect (ZAAS-HFM). The geostatistical treatment of data was performed applying block kriging to obtain interpolation maps of the study area.

3.2 Mineralogical and geochemical methods

Mineralogical characterization of borehole and soil samples was performed by X-ray diffraction (XRD) using a Philips X'Pert powder device with a Cu anticathode and standard conditions: speed 2° 2θ/min between 2° and 70° at 40 mA and 45 kV. The whole sample was examined by crystalline nonoriented powder diffraction on a side-loading sample holder. Semi-quantitative results were obtained by the normalized reference intensity ratio (RIR) method. The mineralogy of the samples was also studied by environmental scanning electron microscopy (ESEM), coupled with energy dispersive X-ray analysis (EDX), using a Philips XL30 microscope. The ESEM was operated at a low-vacuum mode, at a pressure between 0.5 and 0.6 Torr under a water vapor atmosphere and an operating voltage of 20 kV. The XRD and ESEM-EDX analyses were performed at the Centro de Apoyo Tecnológico (CAT Universidad Rey Juan Carlos, Móstoles, Spain). From the total list of major, minor, and trace elements analyzed, Ag, As, Cd, Co, Cu, Fe, Hg, Ni, Pb, S, Sb, Sn, and Zn were specially chosen because of their abundance in these types of sludges and because most of them are included in the priority contaminant list of environmental protection agencies. They were analyzed by total digestion (TD) or lithium metaborate/tetraborate fusion (FUS), inductively coupled plasma-mass spectrometry (ICP-MS), and instrumental neutron activation analysis (INAA) at the Activation Laboratories Ltd. (1428 Sandhill Drive, Ancaster, Ontario, Canada). Quality control at the Actlabs laboratories is performed by analyzing duplicate samples and blanks to check the precision, whereas accuracy is determined using Certified Reference Materials (GXR series; see [15]). Detection limits for the analyzed elements are as follows (data in $\mu g\ g^{-1}$): Ag (0.3), As (5), Cd (0.5), Co (1), Cu (1), Fe (100), Hg (0.005), Ni (1), Pb (5), S (10), Sb (0.5), Sn (1), and Zn (1). Pb content higher than 5000 $\mu g\ g^{-1}$ (above the ICP-MS maximum detection limits) was measured by ICP-OES or atomic absorption.

Water samples were analyzed by ICP-MS at Activation Laboratories Ltd. The pH was measured using an electronic pH meter (CRISON) that was calibrated using standard buffer solutions at two points: pH: 7 and pH: 4. This parameter was determined in a slurry system with an air-dried sample (10 g) mixed with distilled water (25 mL). Before reading the pH values, these solutions were vigorously stirred in a mechanical shaker for 10 min and left to stand for 30 min.

3.3 Complementary techniques

Electrical resistivity tomography (ERT) imaging is a near surface nondestructive technique designed to be widely used in many different geological applications, including the determination of the materials constituting the bedrock, unraveling the stratigraphical record of the basins and locating hidden faults, among others [16]. A resistivity profile is obtained from many different measurements using different available electrode arrays, with the data acquisition being controlled by means of a computer. These measurements provide data about the variations of apparent resistivity values at different depths, in such a manner that when the spacing between the electrodes increases, the resistivity data correspond to a greater depth of investigation. After data acquisition, the apparent resistivity values are converted to an image of true resistivity variations against depth. The resistivity meter used to obtain the data for this study was a Syscal Junior Switch 48. As mentioned before, different electrode arrays are available, with differences in relation to the depth of investigation and signal-to-noise ratio (e.g. [17]). From the different electrode arrays available, a Wenner-Schlumberger array has been selected because it provides a good penetration depth, the signal to noise ratio is good, and both vertical and horizontal resolutions are also reasonable. Moreover, different authors have previously used this array successfully in several similar studies [3–7, 18] because it shows a high contrast between the resistivity values of the vase of the mine ponds and the resistivity values of the infilling. From the field data, the information obtained about the resistance measurements between the different electrodes and distances between them is used to calculate the apparent resistivity values. Then, a plot of the apparent resistivity values vs. depth, named pseudosection, is constructed. Previously to be interpreted, the pseudosections need to be converted into profiles where true resistivity values are plotted against depth. The conversion from apparent to true resistivity values is performed by means of the RES2DINV code. As a first step of this inversion procedure, the data are filtered to remove bad data points, and then the topography information along the profile is also included. The code uses the L1 norm for the data misfit and the inversion was performed using the L1 norm (robust) for the model roughness filter [19]. The choice of the robust inversion is justified because this kind of inversion is more accurate when sharp boundaries in the model exist, and this is just the case involved in this study because of the large contrasts expected in the electrical properties of the materials. The method uses a finite element scheme for solving the 2-D forward problem and blocky inversion method for inverting the ERT data. The code RES2DINV finally provides an inverted resistivity image for each profile. The inverted profile is the one used to obtain the final interpretation about the variations of the subsurface lithology.

Aerial photographs, georeferenced and integrated into a GIS, are used to map morphologic elements and to detect the main changes in the environment through time. The evolution of mine deposits in the Brunita site was carried out to estimate anthropogenic changes in the landscape [5]. We worked with aerial photographs taken in 1929 (photogrammetric flight by Ruiz de Alda), 1946, and 1956 (flights by the Geographic Service of the Spanish Army), and 1973, 1981, 2004, and 2013 (flights by the Spanish National Geographic Institute), and anaglyphs of orthoimages from years 1946, 1956, 1981, and 2004 [20]. On the other hand, the study of a mine pond and a tailing sand dune at the San Quintín mine permitted to evidence the eolian dispersion of contaminants to the surroundings [21]. Several aerial photographs corresponding to different years were also analyzed (1957, Geographic Service of the Spanish army, 1977, 1984, Spanish National Geographic Institute, 2006, digital orthophoto IGN).

4. Geoenvironmental characterization of sulfide mine tailings

The results of the mineralogical and geochemical characterization of the samples collected from tailings, soils, air, water, and watercourse sediments are presented and discussed here. Morphological evolution over time from Brunita and San Quintín mine ponds is presented too, as well as the geophysical study concerning the structure and infilling of ponds and the possible presence of acidic water flows.

4.1 Mineralogical characterization

The mineralogical composition of tailings, colluvial, watercourse sediments, and soil samples has been inferred from the X-ray diffraction studies (**Table 2**). The following nomenclature has been used for tailing-mineral identification: primary minerals, those minerals that constitute ore and gangue assemblages originally deposited in the waste dumps, and secondary minerals, those deposited within the dumps by precipitation from metal-rich waters derived from acid mine drainage.

The nearly homogeneous mineralogical composition of mine tailings is mainly composed of primary gangue minerals from the volcanic or metamorphic host rocks: quartz (30–85 wt%), illite (5–15 wt%), feldspar (5–10 wt%), and chlorite (5–10 wt%). Minor gangue minerals appear in important amounts in some of the areas: siderite (15 wt%) in Brunita. The most important feature of the mineralogical composition of these deposits is the metallic ore mineral contents (25–40 wt%). Significant amounts of pyrite (10–35 wt%), sphalerite (5–10 wt%), and/or galena (5–10 wt%) have been identified in mine tailings (**Table 2**). These high values are probably related to

Mine district	Iberian Pyrite Belt			Cartagena-La Unión	Alcudia Valley		Mazarrón			
Study area	**LN**	**MR**	**MC**	**BR**	**SQ**	**SQ**	**SC**	**SC**	**LM**	**LM**
Sample	**Ta**	**Ta**	**Ta**	**Ta**	**Ta**	**Co**	**Ta**	**WT**	**Ta**	**WT**
Quartz	85	35	35	30	70	80	40	45	50	40
Illite	—	10	5	5	15	10	—	—	—	20
Feldspar	—	10	5	—	—	—	5	30	5	5
Chlorite	5	5	5	10	10	10	—	—	5	10
Calcite	—	—	—	—	—	—	—	—	—	20
Siderite	—	—	—	15	—	—	—	—	—	—
Pyrite	—	25	35	30	—	—	10	5	15	—
Sphalerite	—	10	5	5	—	—	5	5	10	—
Galena	—	—	—	—	—	—	10	5	5	—
Arsenopyrite	—	—	5	—	—	—	—	—	—	—
Magnetite	—	—	5	—	—	—	—	—	—	—
Jarosite	5	5	—	—	—	—	15	5	5	—
Rozenite	—	—	5	—	—	—	—	—	—	—
Goethite	5	—	—	—	—	—	—	—	—	—
Hematite	—	—	—	—	—	—	5	—	5	—
Alunite	—	—	—	—	—	—	—	5	—	—
Gypsum	—	—	5	5	5	—	10	—	—	5

La Naya (LN), Monterromero (MR), Mina Concepción (MC), Brunita (BR), San Quintín (SQ), San Cristóbal (SC), and Las Moreras (LM). Ta: tailings; Co: colluvial; and WS: watercourse sediments.

Table 2.
Semi-quantitative mineralogical composition (wt%) of the studied samples.

inefficient metallurgical processing of the benefited ore during the operational years. Because of the re-working of tailing mine areas, San Quintín area shows the lower ore mineral content. In other cases, like Brunita deposit, different ore minerals amounts are associated with the two different mines exploited and dumped: Brunita and Eloy mines. Cinnabar was identified by X-ray diffraction in one borehole sample from San Quintín. Its presence is due to the experimental metallurgical works carried out during the last period of operations in the Almadén mine (Ciudad Real, Spain). Secondary mineralogy is mainly represented by Fe-sulfates (jarosite and rozenite), Ca-sulfates (gypsum), and Al-sulfates (alunite). Fe-bearing sulfide oxidation increases the metal mobility from these materials compared to the levels mainly composed by sphalerite and galena. Significant amounts of secondary gypsum are typically found in this type of sulfide tailings. Fe-oxides and Fe-hydroxides have also been identified.

In some occasions, the ore minerals are not identified by X-ray diffraction, or are identified in low amounts. In these cases, a detailed study by environmental scanning electron microscopy (ESEM) coupled with energy dispersive X-ray analysis (EDX) is necessary. Four examples of the application of ESEM-EDX are presented in **Figure 2**. Primary sulfide minerals (e.g. galena) identified in low amounts by XRD were also recognized by ESEM-EDX in Monte Romero tailings. Galena occurred as cubic crystals commonly showing octahedron faces. Other sulfide phases such as arsenopyrite, chalcopyrite, and galena were not detected by X-ray diffraction in La Naya tailings. Secondary mineral phases recognized by ESEM-EDX were Fe-oxyhydroxides. Cryptocrystalline Fe-oxyhydroxides frequently occurred around other minerals such as quartz, completely or partially replacing primary sulfides (pyrite and sphalerite). In San Quintín mine, primary ore minerals were not identified by XRD due to the optimized mining works. Pyrite, galena, chalcopyrite, and gangue minerals (barite) were identified by ESEM-EDX (**Figure 2**). In San

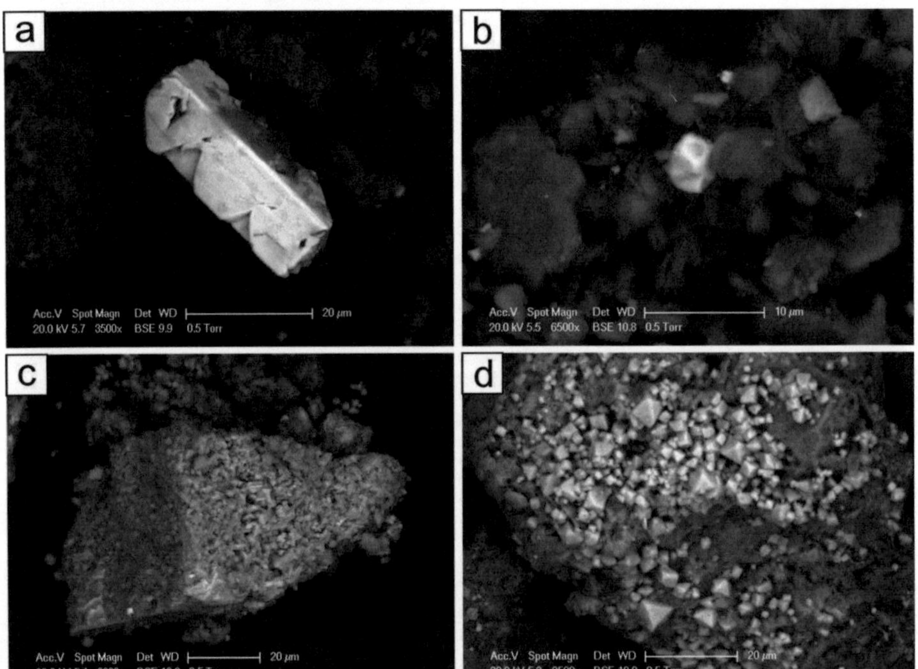

Figure 2.
Backscattered electron (BSE) images: (a) galena crystal from Monte Romero; (b) pyrite crystal from San Quintín; (c) altered faces of a pyrite crystal from San Cristóbal; (d) subidiomorphic magnetite from Las Moreras.

Cristóbal tailings, ore and gangue minerals were identified, as well as oxide minerals such as goethite. In Las Moreras samples, low quantities of Fe-oxides (hematite and magnetite) and carbonates (siderite) were identified by ESEM.

With respect to the colluvial sediments drilled at the San Quintín boreholes, the mineralogical composition was totally composed by primary silicates, as well as watercourse sediments from Las Moreras area (**Table 2**). The ore mineral content is low enough to be identified by XRD analysis. In contrast, the semi-quantitative mineralogical composition of watercourse sediment from San Cristóbal mine included ore minerals (pyrite, sphalerite, and galena) and secondary sulfates (jarosite and alunite).

4.2 Geochemical characterization

4.2.1 Mine tailings

Total ferric iron (Fe_2O_{3total}), S, and trace element (Ag, As, Au, Cd, Cu, Ni, Pb, Sb, Sn, and Zn) concentrations, and pH values from tailing samples of the four mine district are summarized in **Table 3**. All samples showed a pH range of 2.2–5.6. This value range reflects the typical acid character of stored mine tailings. The composition of all tailing samples is characterized by the high contents of ore-bearing elements in each district: As, Cu, and Pb in the Iberian Pyrite Belt, Pb and Zn in Cartagena-La Unión and Alcudia Valley, and As, Pb, and Zn in Mazarrón. The total ferric iron content is significantly high in analyzed samples from all mine districts due to the omnipresence of Fe-bearing minerals like pyrite. The significantly high contents of potentially hazardous elements like Fe, Cu, Pb, and/or Zn are due to the nature of the mined ore, which is mainly composed of pyrite, chalcopyrite, sphalerite, and galena (**Table 2**). The highest metal contents are related to the mining history of each district, and the efficiency of the metallurgical processing in the benefited ore during the operational period of time. In the case of San Quintín mine, approximately 3 million tons of minerals from the tailings were re-worked. Then, the lowest Pb and Zn contents are located at the upper levels of the ponds. High Hg content measured in the San Quintín mine tailings is related to the experimental metallurgical works previously cited (Section 2). Significant Hg values were measured in Monte Romero mine related to the formation of a replacive mineralization. Pb (up to 21,130 $\mu g\ g^{-1}$) and Zn (41,841 $\mu g\ g^{-1}$) contents in the tailings from Mazarrón district (**Table 3**) as well as the significant Ag content from San Cristóbal mine related to the exploitation of Ag-bearing galena deserve special mention.

The Mina Concepción samples were collected with a manual sampler from the first meter in depth. That is the reason for the lower metal contents to be associated with the more recent and efficient metallurgical works. Related to the Iberian Pyrite Belt district, relevant variations as a function of depth were identified in all of the analyzed element contents from Monte Romero samples. Possible explanations for these variations could be argued: (a) periods with higher mineral benefit, due to improvements in metallurgic processes or to a higher grade mineralogy and (b) a change in the exploitation targets, originally focused on galena (Pb) mining but later re-directed to pyrite (Fe) and sphalerite (Zn) mining due to environmental policies that do not recommend the use of lead in many industrial fields.

4.2.2 Sediments and soils

Total ferric iron, S and trace element concentrations, and pH values from colluvial and watercourse sediment, and soil samples of the Alcudia Valley and Mazarrón districts are summarized in **Table 4**.

Mine district	**Iberian Pyrite Belt**			**Cartagena-La Unión**	**Alcudia Valley**	**Mazarrón**	
Study area	**LN**	**MR**	**MC**	**BR**	**SQ**	**SC**	**LM**
Ag	b.d.	31–81 (48)	b.d.	2–8 (4)	5–29 (12)	20- > 100 (55)	14–26 (20)
As	191–909 (472)	1110–2740 (1650)	223–1080 (613)	39–385 (195)	16–54 (28)	400–633 (490)	181–630 (366)
Au	49–100 (70)	311–744 (558)	22–86 (53)	n.a.	n.a.	n.a.	n.a.
Cd	n.a.	n.a.	n.a.	3–67 (22)	5–22 (13)	4–833 (110)	97–373 (186)
Cu	306–4511 (998)	914–16,582 (4874)	n.a.	76–323 (179)	42–381 (171)	121–882 (406)	168–356 (230)
$Fe_2O_{3\,total}$	8.51–14.1 (11.5)	2.12–27.2 (14.57)	6.11–24.1 (14)	22.3–52.6 (38.7)	4.5–6.3 (5.6)	9.8–22.9 (19.2)	17.7–28.5 (23.7)
Ni	b.d.	b.d.	b.d.	11–40 (22)	29–61 (44)	11–41 (21)	41–59 (49)
Pb	51–222 (133)	295–12,610 (4615)	71–475 (264)	1610–5950 (3211)	1510–10,500 (3992)	5987–39,877 (21,130)	3940–5239 (4665)
S	n.a.	n.a.	n.a.	3.8–19.1 (11.8)	0.3–1.0 (0.4)	3.5–53.0 (6.6)	7.1–12.8 (9.3)
Sb	31.3–50.4 (35.2)	168–861 (338)	16.2–67.5 (40)	3–54 (30)	36–162 (79)	108- > 200 (149)	54–136 (93)
Sn	b.d.	b.d.	b.d.	14–244 (76)	3–7 (5)	8–121 (83)	24–37 (30)
Zn	50–260 (143)	0.5–6.9 (4)	200–970 (511)	2020–2112,150 (6315)	1250–4470 (2418)	2500–11,405 (4101)	12,810–41,841 (27,738)
pH	2.8–3.5 (3.2)	2.5–3.2 (3.0)	n.a.	2.4–3.7 (3.0)	n.a.	2.2–3.6 (2.5)	2.8–5.6 (4.0)

Ag, As, Au, Cd, Cu, Ni, Pb, Sb, Sn, and Zn in μg/g. Fe_2O_{3total} and S in wt%. Mean in brackets. b.d.: below detection; n.a.: not analyzed. La Naya (LN), Monterromero (MR), Mina Concepción (MC), Brunita (BR), San Quintín (SQ), San Cristóbal (SC), and Las Moreras (LM).

Table 3.
$Fe_2O_{3\,total}$, trace element content, and pH values in the studied tailings.

Mine district	**Alcudia Valley**			**Mazarrón**	
Study mine	**San Quintín**			**San Cristóbal**	**Las Moreras**
Sample	**Colluvial**	**Soil**	**Blank**	**Sediment**	**Sediment**
Ag	0–3 (2)	0–3 (1)	0.2	21–60 (38)	0–7 (3)
As	8–24 (17)	b.d.-26 (15)	11	216–312 (259)	15–131 (33)
Cd	1–7 (3)	b.d.-7 (b.d.)	b.d.	2–4 (3)	1–6 (3)
Cu	38–196 (76)	5–39 (17)	18	70–122 (96)	45–482 (177)
$Fe_2O_{3\ total}$	5.3–9.5 (6.9)	1.7–4.7 (3.5)	4.2	7.6–33.5 (18.3)	4.0–12.0 (5.9)
Ni	46–93 (59)	13–46 (34)	30	10–14 (12)	30–61 (47)
Pb	79–577 (315)	41–1110 (318)	34	9395–16,193 (12,800)	157–1880 (555)
S	0–0.2 (0.1)	0–0.3 (0.1)	0.01	2.7–4.5 (3.9)	0.1–3.0 (1.0)
Sb	5–40 (24)	2–30 (9)	2	98–124 (111)	2–48 (14)
Sn	3–5 (4)	2–3 (3)	2	5–10 (7)	2–6 (4)
Zn	186–844 (503)	34–1180 (335)	49	815–1660 (1334)	452–10,693 (2907)
pH	n.a.	5.7–6.2 (5.9)	5.1	3–3.4 (3.2)	7.0–7.7 (7.4)

Ag, As, Au, Cd, Cu, Ni, Pb, Sb, Sn, and Zn in μg/g. Fe_2O_3 and S in wt%. Mean in brackets. b.d.: below detection. n.a.: not analyzed.

Table 4.
$Fe_2O_{3\ total}$, trace element content, and pH values in the studied colluvial, soil, and watercourse sediment samples.

The highest contents were found in the pond samples, and the intermediate contents in the colluvial samples from San Quintín area. The ponds were not waterproofed, and hazardous metals from the upper ponds have percolated through the underlying colluvial sediments. Cu, Pb, and Zn contents show significant amounts. Five representative soil samples were analyzed in order to determine the importance of contamination (**Table 4**). Two mine soil samples show similar metal and As content to the upper tailing samples. The other two were agricultural soil samples, showing significantly lower metal and As content, but higher Hg and Pb contents than the local background sample (blank in **Table 4**), collected from an agricultural soil 4.5 km to the south-east. However, remarkably high As, Pb, and Zn contents were still found in this background sample, suggesting that the surrounding agricultural soils are also contaminated. In fact, Ag, Cd, Pb, and Zn contents from agricultural soil samples are higher than geochemical baselines reported by [22] for this Spanish region.

In Mazarrón district, both tailings and watercourse sediments showed high amounts of potentially toxic elements, slightly lower at the sedimentary level (3.0–5.5 m depth). The total iron content ranged between 4.0 and 33.5 wt%, Pb was 157–16,193 $\mu g\ g^{-1}$, and the Zn content ranged between 815 and 10,693 $\mu g\ g^{-1}$ (**Table 4**). Other trace elements that displayed high values were: As, Cu, and Sb. The sediments mark a defined geochemical limit with the tailing unit. The upper mine tailings are significantly concentrated in Fe_2O_{3total} and heavy metals, whereas the sediments display marked lower values. This decrease is not absolutely regular, with major peaks in Cu and Zn contents, and minor increases in As, Cu, Fe_2O_{3total}, Pb, and Zn. The amount of calcite in the Las Moreras sedimentary unit (**Table 2**) controls the pH, buffering to within a small range of 7.2–7.7 (**Table 4**). In turn, the

Mine district	Iberian pyrite belt	Cartagena-La Unión		Alcudia Valley		Mazarrón	
Study mine	MC	BR		SQ		SC	LM
Sampling	LK (5)	WT	BO	WT (6)	AMD (3)	WT	WC
As	>2000	79	2.6	>1.5	2.3	>2000	b.d.
Cd	14–54	483	3.8	0.1–26.6	>3200	6470	1.5
Cu	>2000	>2000	12.3	1–12.6	>8300	>2000	52
$Fe_2O_{3\,total}$	>100,000	>100,000	90	30–100	>189,000	>100,000	100
Ni	30–143	559	80.1	1.8–22.3	>3500	5620	120
Pb	0–67	199	8.5	0.4–21.6	>2800	0.5	1
Sb	0–1	b.d.	6.1	0.1–3.0	1.3	11.8	1.4
Sn	b.d.	b.d.	b.d.	b.d.	b.d.	3	2
Zn	>2500	>2500	93.6	57–2520	>550,000	>2500	365
pH	2–2.6	2.4	5.7	6.2–7.1	2.5–4.3	1.8	8.3

Values in μg/L. b.d.: below detection. Mina Concepción (MC), Brunita (BR), San Quintín (SQ), San Cristóbal (SC), and Las Moreras (LM). LK: leakage; WT: water table; BO: borehole; AMD: acid mine drainage; and WC: watercourse.

Table 5.
$Fe_2O_{3\,total}$, trace element content, and pH values in the water samples.

upper tailing unit of Las Moreras shows much lower pH values (2.8–5.6), due to the sulfide content and the complete absence of calcite.

4.2.3 Waters

The composition of leakage sample waters of a restored mine pond (Mina Concepción) indicates that these waters represent acid mine drainage, as reflected by their very low pH (<2.6) (**Table 5**). Trace element contents are very high for Cu and Zn (>2 mg/l), both higher than the EPA's maximum recommended limits for irrigation waters (0.2 mg/l for Cu and 2 mg/l for Zn [23]). This is also the case for As in samples leaking from the dyke wall and the puddle. Lead also goes beyond the legislation limits in samples from the dyke wall and the drainage pipes. Acid mine drainage was observed in the northern part of the Brunita mine pond (pH < 2.4) (**Table 5**). The concentrations were very high for Cu, Zn, Cd, Ni, and Fe in the water sample. Results from complementary techniques (ERT), shown in Section 4.3 of this chapter, have confirmed the formation of AMD waters in Mina Concepción and Brunita mines.

One water sample was collected at 8 m depth in the borehole from San Quintín mine (**Table 5**). The high EC and acidic pH values are consistent with water from ore deposits retained in tailing ponds. Three samples showing low pH and significantly high trace element contents indicate AMD flowing from the remaining tailings. AMD was not observed in samples from the watercourse crossing the mining zone (**Table 5**). pH values in these waters are circumneutral, and EC values and metal contents are significantly lower than in samples from the tailings. Samples collected up- and downstream display the lowest trace element contents. Higher metal contents have been measured in the rest of watercourse samples, denoting that trace element contamination occurs through the mining area.

With regard to the San Cristóbal mine pond, AMD was clearly detected in the water sample: pH < 2, high redox potential, high EC, and Total dissolved salt values. Concentrations of trace elements were very high for As and Cu (>2000 μg/L), Zn

(>2500 μg/L), Cd and Ni (>5600 μg/L), and Fe (>100,000 μg/L). Water from the seasonal watercourse of Las Moreras was also analyzed. Significant contents of metallic elements (Cu, Fe, Ni, and Zn) were measured, all beyond the established limits for irrigation waters.

4.2.4 Hg in air

A singular case occurs in San Quintín mine where significantly high Hg content has been identified in the mine area. Gaseous Hg emissions were measured from the tailings and surrounding soils (**Figure 3**). The total gaseous mercury distribution in the studied area significantly changes between summer and winter. The area affected by TGM values up to 100 ng m^{-3} is restricted to the surroundings of the cinnabar stockpile in winter, but the affected area is 0.16 km^2, and extends into the Don Quixote Route in summer. TGM values are lower than the limit recommended for the general population by the World Health Organization (WHO) (1000 ng m^{-3}) for the worst scenario [24]: higher temperature and solar radiation during summer.

4.2.5 Multivariate analysis

Multivariate analysis has been carried out on the significant metal contents from samples of tailings (Brunita), tailings + colluvial sediments (San Quintín), and tailings + watercourse sediments + bedrock (San Cristóbal and Las Moreras) (**Figure 4**). Statistical data processing was carried out using Minitab® 16 software. The multivariate analysis was based on clustering (group average linkage dendrograms, Euclidean distance) of the set of samples and significant trace elements (Ag, As, Cd, Cu, Pb, Sb, Sn, and Zn). The dendrogram of the metals and As in the Brunita tailing samples shows the metallic signature of the district ores: Ag-Pb-Cd-Zn, Cu, and As-Sb-Sn, with As being mainly related to Sb (tetrahedrite-tenanntite mineral group). Ag-Pb-Cd-Zn signature is clearly defined due to the mineral source. In the case of San Quintín, the dendrogram from tailings and colluvial sediments reflects again the metallic signature of the district (Pb-Ag-Sb, Cu, and Zn-Cd to a certain extent [13]), with As mainly related to Sb (bournonite and boulangerite) and Pb-Ag (galena). Some samples display a strong affinity to the Ag-Pb-Sb-As association, whereas other samples display Cd-Zn affinity. The same

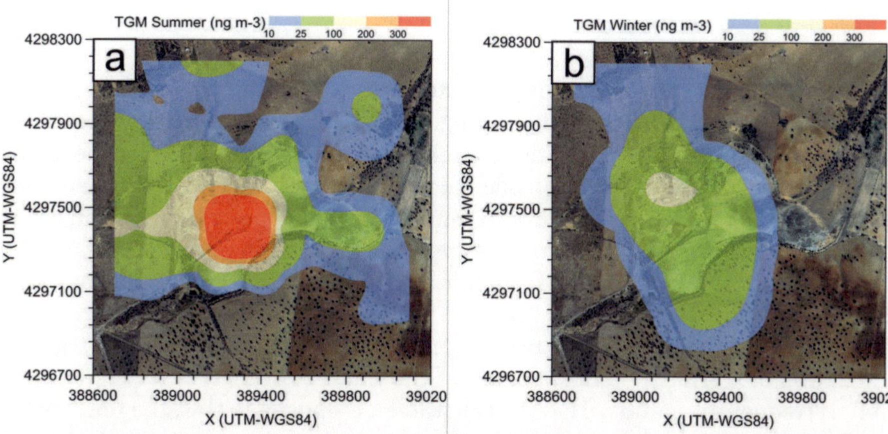

Figure 3.
Total gaseous mercury (TGM) seasonal distribution in the San Quintín area: (a) summer and (b) winter values. Modified from Martín-Crespo et al. [6].

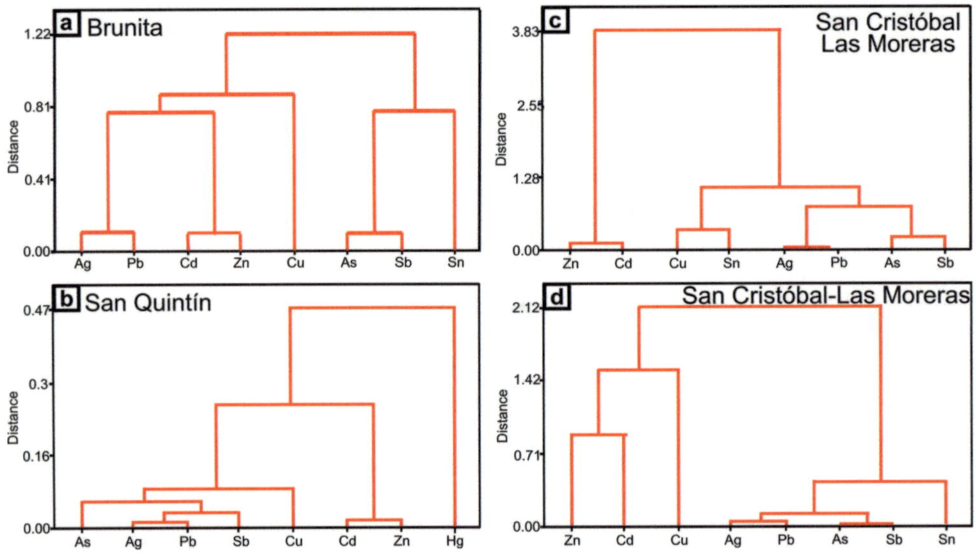

Figure 4.
Dendrograms (distance Euclidean) of metals: (a) Brunita tailings; (b) San Quintín tailings and colluvial; (c) Mazarrón tailings; and (d) Mazarrón watercourse sediments and bedrocks.

metallic signature has been obtained from tailings and colluvial sediments, reflecting the same origin for both kinds of samples. The external origin of Hg is reflected by the highest obtained distances. The same occurs for the samples from Mazarrón district, reflecting the metallic signature of the district (Pb-Ag-Sb, Zn-Cd, and Cu to a certain extent [14]) with As being mainly related to the Sb (stibnite and sulfosalts) and Pb and Ag (galena). Some slight differences are displayed in the samples from sediments and bedrocks, particularly the larger range of Cu content.

All dendrograms presented in **Figure 4** are in good agreement with field data, and mineralogical and geochemical features of tailings and watercourse deposits (**Tables 2–4** respectively). In summary, the metallic signature of the three districts is clearly defined in the samples from tailings and affected sediments.

4.3 Complementary techniques

4.3.1 Electrical resistivity tomography (ERT)

Additional information about the characteristics of the mine tailing deposits can be obtained from electrical resistivity tomography (ERT) data. The major pieces of information that this method provides are related to both the thickness of the deposits and the occurrence of AMD (both inside the mine pond as flowing out through the dyke or the base). **Figure 5** shows several examples of the type of information derived from the application of ERT to different mine ponds, resulting in a valuable tool that completes the information derived from mineralogical and geochemical techniques.

As a general rule, the materials that constitute the mine pond infilling are characterized by a medium to fine texture and high-water content. Moreover, due to oxidization of sulfide minerals, the pH of the water stored in the mine pond infilling is frequently acidic (pH < 5) in character. Opposite to this, the host rock where the mine pond is placed should be the host rock of the mineralization, typically metamorphic and/or igneous rocks of coarse texture and extremely low water content. Thus, a high resistivity contrast between the mine pond infilling (low to very low resistivity values) and the host rock (medium to high resistivity values)

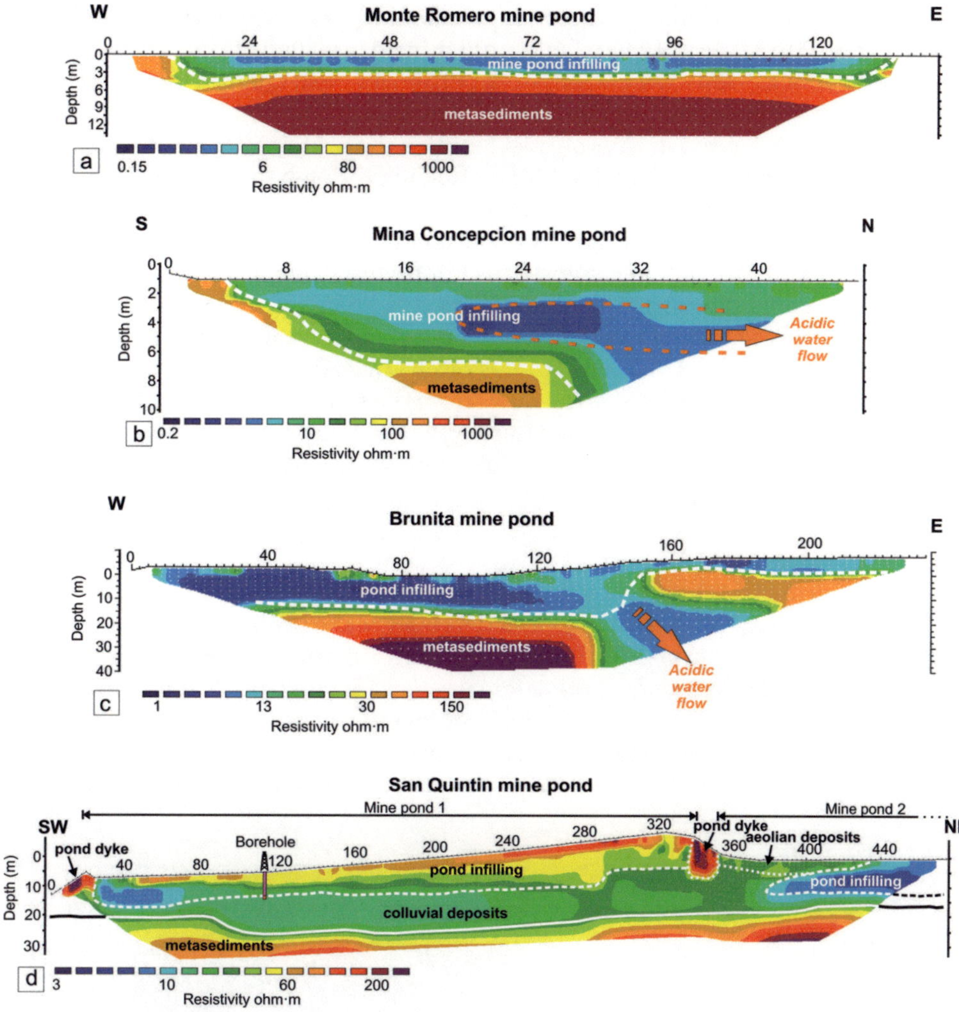

Figure 5.
ERT profiles obtained at four different mine sites. Each profile provides information about the thickness of the mine deposits and the presence or not of acidic water.

exists, allowing an accurate characterization of the boundary between both rock types and providing good estimations of the thickness of the mine pond deposits. Different mine pond thickness values and bottom geometries (dashed white lines) are imaged in **Figure 5**. Monte Romero and San Quintin mine ponds show simple bowl-shaped geometries with a thickness of ~3 and 10 m, respectively (confirmed with data from a borehole in the case of San Quintin mine pond 1), whereas Mina Concepcion and Brunite mine ponds exhibit a stepped bottom geometry with variable thickness (~6–10, and ~5–12 m, respectively). Where different rock units are present below the mine pond, instead of a homogeneous lithology, an estimation of the thickness of the different units can also be obtained. This is the case for San Quintin mine ponds, where a ~10 m thick sedimentary unit of colluvial deposits overlies the metasediments that constitute the regional basement. As mentioned before, a low pH value for the water contained in the mine pond deposits is also frequent, resulting in lower resistivity values in comparison with water with circumneutral pH. Therefore, the occurrence of acidic water inside a mine pond is revealed by extremely low resistivity values, normally lower than 1 ohm m. This is the case for Monte Romero, Mina Concepcion, and Brunita mine ponds where areas of <1 ohm m inside the mine pond correspond to the presence of water with pH ranging from 2 to 3 (see **Table 5**). On the other hand, the higher (>5 ohm m, and mainly >10 ohm m) resistivity values of the infilling of San Quintin mine ponds are associated with circumneutral pH (**Table 5**).

Finally, the strong resistivity contrast between the acidic water and the host rock results to be very useful to detect if AMD is flowing through the bottom of the mine pond. Where the sealing of the mine pond is correct, the host rock shows homogenous high resistivity values along the whole boundary with the pond infilling, such as the case of Monte Romero and San Quintin mine ponds. However, where AMD flows through the host rock, discrete areas of resistivity values much lower than the ones associated with the host rock are imaged, revealing the occurrence and sense of flow of the AMD. The latter is nicely imaged in both the cases of Mina Concepcion mine pond, where AMD flows from the inner central part of the pond toward the northern edge (confirmed during the field inspection of the dyke that exhibits AMD trough it), and Brunita mine pond, where AMD flows toward the east through the host rock.

4.3.2 Aerial photographs

Dumping of huge piles of tailings and debris since the beginning of mining operations at the Brunita area leads to a deep transformation of the landscape with the whole disappearance of the original reliefs [5] (**Figure 6**). A few studies are recognized in the 1929 and 1946 aerial photographs, but mine ponds were not yet operational. The orography consisted of smooth hills (~40 height difference) separated by NNW–SSE and WNW-ESE ravines (<500 m long). However, by 1956,

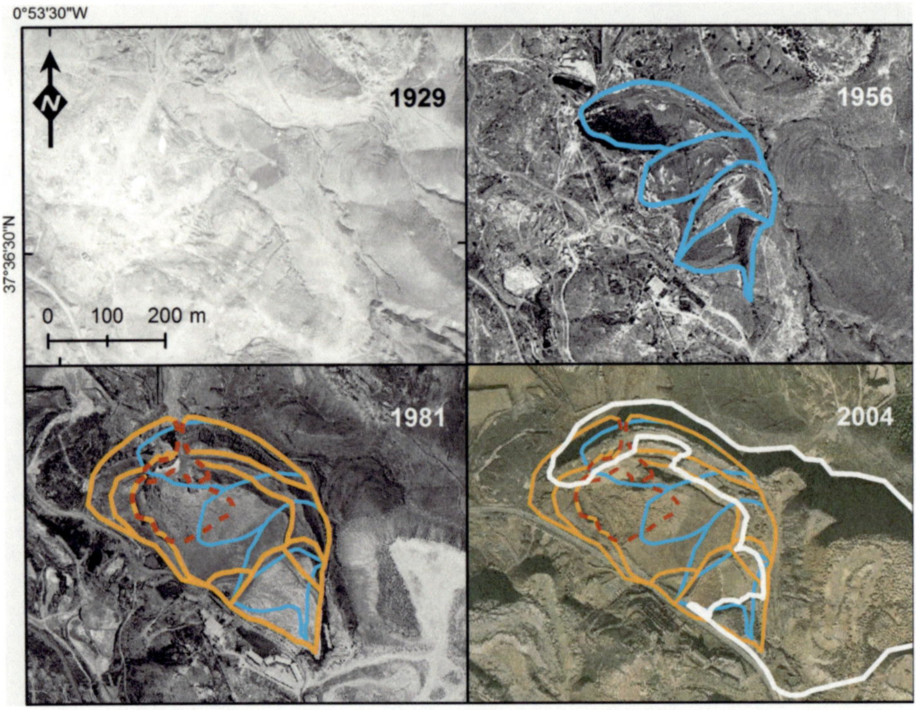

Figure 6.
Landscape evolution in the surroundings of the Brunita mine pond from aerial photographs taken in: 1929 (Ruiz de Alda photogrammetric flight); 1956 (Geographic Service of the Spanish Army); and 1981 and 2004 (Spanish National Geographic Institute). Blue lines, mine ponds in 1956; orange lines, mine ponds in 1981; brown dashed lines, badlands and landslide scar in 1981; and white line, spoil tips in 2004.

tailings from the mineral treatment plant were dumped in four stepped ponds along a NNW–SSE valley. Due to the continued mine activity, the tailings leveled the land up to the highest pond between 1973 and 1981, even burying the adjacent hills, and debris from the mine quarry began to cover the easternmost NNW–SSE ravines and hills. At this stage, there were two ponds, a large one with three stepped dikes and a small one in the highest part of the valley. The landslide scar in the dikes and the back sunken area, due to the flowage of tailings that caused the disastrous 1972 flash flood, are also visible. The sunken area was crisscrossed by gullies due to subsequent water erosion. Although the mine was closed in 1981, the reinforcement of the pond perimeter, with the sealing of the broken area, and the strong accumulation of debris to the east completely buried the ravines and hills and leveled the topography by 2004. Changes in the gully drainage pattern and the retreat of scarps indicate that tailing erosion persists.

Eolian dispersion of contaminants from a tailing sand dune (**Figure 1**) is the most important environmental concern at the San Quintín area [21]. By 1977, the mine pond was divided into three sectors through dikes. One of these dikes will be the obstacle over which the dune will be developed when the mine operations ceased. The dune has been growing and migrating by the dominant winds since 1984. As the tailings are not replaced, the dune is losing its pollutant particles by dispersion toward the nearby river and agricultural soils, so its decrease in size or disappearance is expected.

5. Environmental concern

The results obtained from the mineralogical and geochemical characterization of the samples collected from tailings, soils, air, water, and watercourse sediments allow identifying the potential environmental concerns that would affect the different mine districts. These potential environmental concerns can be classified according to three main types: (a) ecosystem risks, (b) human health risks, and (c) physical hazards. Ecosystem risks are mainly related to the negative effect of both the acidic water and metals. To properly evaluate the potential volume of metals susceptible to produce negative effects on the ecosystems, the mineralogical and geochemical characterization of the tailings is crucial. From the cartographic (area) and ERT (general geometry and thickness) studies, an infilling volume of 912,000 m^3 has been calculated for the Brunita mine pond. The maximum amounts of potential contaminants were obtained taking into account the mean content of potentially toxic elements (**Table 3**), the previously calculated volume, and the mass of the waste. A mean bulk density of 2.65 g/cm^3 was calculated from the mineral particle density and assuming a porosity of 40%, which is the value for mine ponds originating from the processing of this type of deposit. From the mean trace element content shown in **Table 3**, the Brunita impoundment contains more than 24,250 t of potentially toxic elements such as (470 t), Cd (52 t), Cu (430 t), Ni (53 t), Pb (7753 t), Sb (71 t), Sn (184 t), and Zn (15,245 t). Release of these amounts of toxic elements would be catastrophic for the environment and the community (death, serious material damage, coastal areas, and farm land). Similar studies in the Iberian Pyrite Belt district show amounts of potentially toxic elements of 5900 t in La Naya, and 2100 t in Monte Romero ponds.

In order to evaluate the contaminating degree of tailings, the geo-accumulation index (I_{geo}) was calculated. Müller [25] defined I_{geo} and enabled the assessment of sediment contamination by comparing current and pre-industrial concentrations of heavy metals. This index is mathematically expressed as $I_{geo} = \log_2 C_n/1.5B_n$, where C_n is the concentration of an element in the sample and B_n is the background concentration of the corresponding element in the Earth's crust, according to [26]. Müller [25] suggested six descriptive classes for this index: uncontaminated ($I_{geo} \leq 0$), uncontaminated to moderately contaminated ($0 < I_{geo} < 1$), moderately contaminated ($1 < I_{geo} < 2$), moderately to strongly contaminated ($2 < I_{geo} < 3$), strongly contaminated ($3 < I_{geo} < 4$), strongly to extremely contaminated

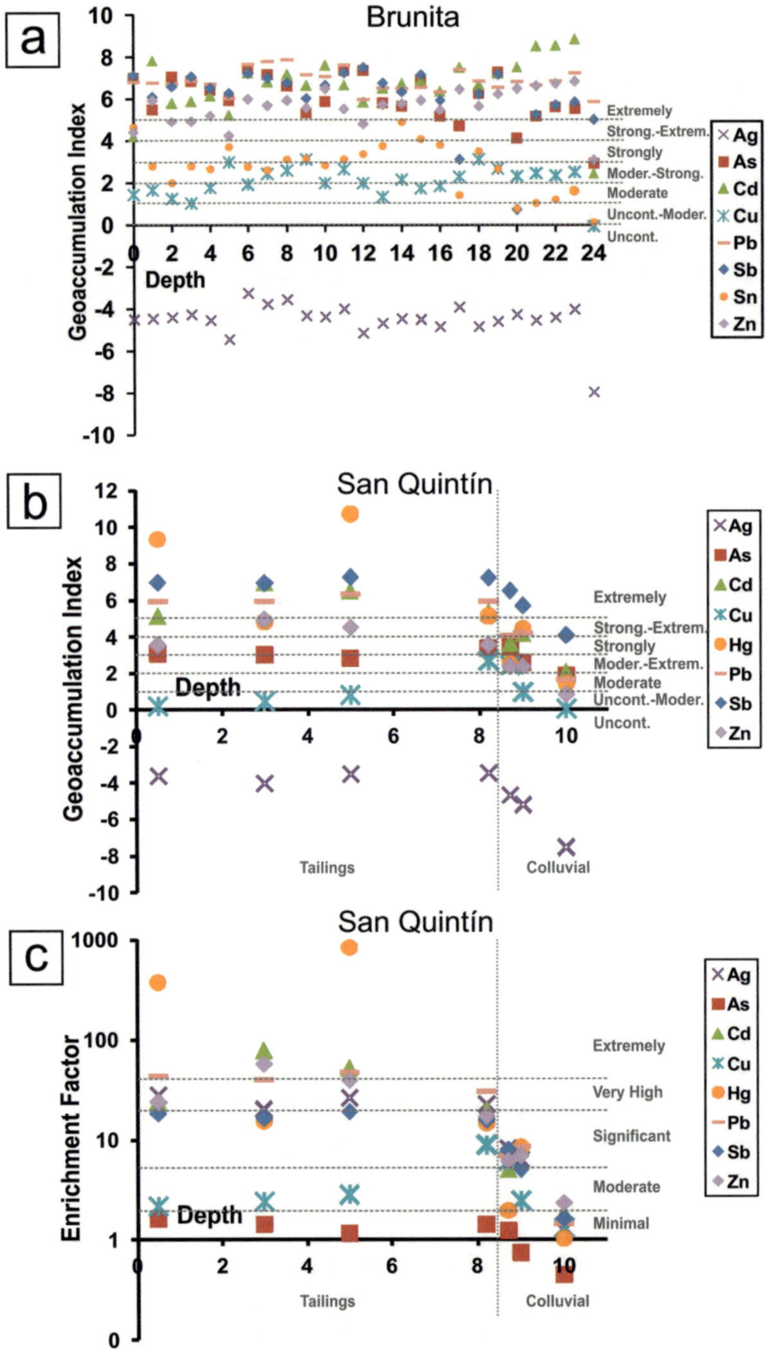

Figure 7.
(a) Geoaccumulation index for Brunita tailings, (b) geoaccumulation index for San Quintín tailings and colluvial; (c) enrichment factor for San Quintín tailings and colluvial. Modified from Martín-Crespo et al. [5, 6].

$(4 < I_{geo} < 5)$, and extremely contaminated $(I_{geo} > 5)$. The I_{geo} index was calculated for tailings from Brunita, and tailings and colluvial from San Quintín (**Figure 7**). As, Cd, Pb, Sb, and Zn from Brunita tailings show extreme contamination $(I_{geo} > 5)$, whereas Cu and Sn show moderate to strong contamination $(1 < I_{geo} < 4)$. Ag is classified as a nonpollutant. The contamination classes are two levels higher than those obtained for similar tailings in Spain [6]. Cd, Hg, Pb, and Sb show extreme contamination $(I_{geo} > 5)$, and As and Zn show moderate to heavy contamination $(1 < I_{geo} < 5)$ in the tailings and colluvial sediment from San Quintín. Cu shows moderate to heavy contamination, and Ag is classified as unpolluted (**Figure 7**). Sutherland [27] proposed the enrichment factor (EF) to assess the level of contamination and the possible anthropogenic impact. To identify anomalous metal concentration, geochemical normalization of the heavy metal data to a conservative element, such as Fe, was employed (geochemical normalization). EF was calculated using the formula $EF = (M/Fe)_{sample}/(M/Fe)_{background}$, where $(M/Fe)_{sample}$ is the ratio of metal to Fe concentrations in the sample and $(M/Fe)_{background}$ is the ratio of metal to Fe concentrations of the background (blank; **Table 4**). Sutherland [27] proposed five contamination categories: minimal enrichment (EF < 2), moderate enrichment (2 < EF < 5), significant enrichment (5 < EF < 20), very high enrichment (20 < EF < 40), and extremely high enrichment (EF > 40). The San Quintín samples show very high to extremely high enrichment in Ag, Cd, Hg, Pb, Sb, and Zn. EF values for As are significantly lower than for the rest of elements, reflecting the lack of As-bearing minerals. **Figure 8** shows I_{geo} and EF for San Quintín representative soil samples. Agricultural soil samples (S-06 and S-53) and the background sample (S-00) show the same features: they are moderately contaminated by As, Cd, Pb, and Sb and not contaminated by Ag, Cu, and Zn. The I_{geo} for Hg was strong for agricultural soils and extreme for mine soils. Agricultural soil samples (S-06; S-53) and mine soil sample (S-37) show minimal or moderate EF for Ag, As, Cd, Cu, Pb, Sb, and Zn. The EF values for Hg were significant or very high for agricultural soils and extremely high for mine soils. These data highlight the significant metal contents of the mine site, which can become especially hazardous due to eolian dispersion.

The occurrence of AMD inside the tailings and its flow through the mine deposits toward the surrounding environment represents a major risk for the ecosystems. In this sense, several zones have been affected by metal mobilization though acidic water and its percolation from tailings to riverbed deposits, resulting in the affection of watercourses (Mina Concepcion) and groundwater (Brunita). Consequently, Mazarrón and Iberian Pyrite Belt districts show water metal contents beyond the EPA's maximum recommended limits in irrigation waters. Where AMD is confined inside the mine tailings (Monte Romero and San Quintín), metal mobilization also occurs but the affection to the environment is limited. However, the large volume of acidic water with high metal contents stored at these deposits

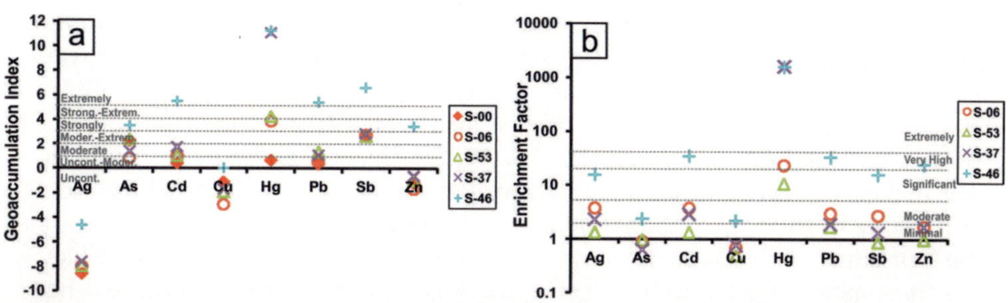

Figure 8.
Representative soils from San Quintín mine area: (a) geoaccumulation index and (b) enrichment factor. Modified from Martín-Crespo et al. [6].

represents a major potential ecosystem risk. If a failure of the dam occurs, or the sealing of the mine pond fails, the ecosystem, watercourses, and riverbed sediments would be largely affected by the release of acidic water and its dissolved hazard metals.

Regarding human health risks, they are mainly associated with the eolian dispersion of contaminants. San Quintin mine ponds represent the area with the higher risk due to the combined effect of both the eolian dispersion of metals from the dune developed on the mine tailings, affecting the surrounding agricultural soils, and the gaseous mercury emissions. As previously mentioned in Section 4.2.2, agricultural soils surrounding San Quintín mine display As, Cd, Pb, and Zn contents higher than geochemical baseline. Therefore, they are contaminated and can be considered as a potential human health risk by the metal input to the olive tree crops. Nevertheless, metal contents in water from the watercourse crossing the mining area are below recommended limits for irrigation waters, denoting not significant affection by AMD. Although this zone is not remediated and not in a condition for public transit, the San Quintín mine has been reported as one of the points to be visited on the longest Eco-tourist Itinerary in Europe, named "Don Quixote Route, a place for adventure". Section four of the route crosses the San Quintín mining area, exhibiting ruinous mine structures. This mine has become a representative example of the socio-economic and cultural benefits that its restoration could confer to this zone. Although these types of tourist initiatives are remarkable in terms of geological heritage, a previous characterization and reclamation study has not been carried out.

Physical hazards are mainly related to the presence of open shafts and unstable ponds and have also been identified in the different mine districts. Alcudia valley and Mazarrón districts contain many abandoned open shafts and tunnels. Unstable ponds have also been identified as in the case of Brunite mine pond, where a previous dam failure occurred [5]. Similar to this, the outflow of acidic water through the dam of Mina Concepcion mine pond would represent a source of instability resulting in a potential physical hazard.

6. Conclusions

This work revealed that the joint use of mineralogical, geochemical, and geophysical techniques can provide an environmental characterization of abandoned mine sites, allowing for estimations of potential pollution and the extent of affected zones.

Significant potentially hazardous element contents have been identified in all studied mine districts, not only in the mine tailings but also in the underlying colluvial and alluvial sediments and surrounding soils. Mineralogical and geochemical signatures of the ore mineralization are clearly recognized in all analyzed samples. Pyrite, sphalerite, and galena are the main ore minerals identified in the mine tailings. Gangue minerals (quartz, illite, feldspar, and chlorite) and secondary minerals (Fe-sulfates, gypsum, and Fe-sulfates) have also been identified by XRD and/or ESEM-EDX. Significantly high contents of As, Cu, Pb, and Zn have been identified in the majority of the mine tailings, reflecting the related environmental hazards associated with all of these abandoned deposits. Moreover, significant potentially toxic element content has been analyzed in tailings from restored mine pond like Mina Concepción. Agricultural soil samples show lower metal and As content but higher Hg and Pb content than in the background sample in the San Quintín area. AMD has been clearly identified not only flowing from the remaining tailings, but also from a restored mine pond, denoting that environmental hazard persists.

ERT provides valuable additional information about the mine deposits. The strong resistivity contrast between the infilling and the underlying rock allows obtaining both the thickness of the infilling as the geometry of the bottom mine pond. Moreover, if the infilling deposits contain water, the resistivity values provide information about both the acidic character of the water and the occurrence or not of AMD flow outside the mine pond. The mapping of mine deposits from time series of aerial images reveals the strong impact of mining on the landscape due to the dumping of large amounts of polluting wastes and their mobilization thereof to the surrounding areas by several geological processes (mass movement, gully erosion, and eolian dispersion).

Major environmental hazards are associated with different main pathways (wind erosion and water flows) and several receptors (bathing waters, agricultural soils, humans, and sediments) depending on the specific mine area. In summary, this type of abandoned deposits need to be characterized, monitored, and restored in order to avoid mobilization of tens thousands of tons of potentially hazardous elements.

Acknowledgements

This work has been accomplished on the frame of projects URJC-RNT-063-1, URJC-CM-2006-CET-0636, and URJC-CM-2008-CET-3644 funded by Comunidad de Madrid and Universidad Rey Juan Carlos. The Open Access Publishing Fee has been funded by the Universidad Rey Juan Carlos.

Conflict of interest

The authors declare no conflict of interest in this chapter.

Author details

Tomás Martín-Crespo*, David Gómez-Ortiz and Silvia Martín-Velázquez
Department of Biology and Geology, Physic and Inorganic Chemistry,
University Rey Juan Carlos, Madrid, Spain

*Address all correspondence to: tomas.martin@urjc.es

References

[1] Lindsay MBJ, Moncur MC, Bain JG, Jambor JL, Ptacek CJ, Blowes DW. Geochemical and mineralogical aspects of sulfide mine tailings. Applied Geochemistry. 2015;**57**:157-177

[2] IGME. Inventario nacional de depósitos de lodos [Internet]. 2002. Available from: http://info.igme.es/balsas/ [Accessed: 2018-11-30]

[3] Martín-Crespo T, Martín-Velázquez S, Gómez-Ortiz D, De Ignacio-San JC, Lillo J. A geochemical and geophysical characterization of sulfide mine ponds at the Iberian Pyrite Belt (Spain). Water, Air, and Soil Pollution. 2011;**217**:387-405

[4] Martín-Crespo T, De Ignacio-San JC, Gómez-Ortiz D, Martín-Velázquez S, Lillo J. Monitoring study of the mine pond reclamation of Mina Concepción, Iberian Pyrite Belt (Spain). Environmental Earth Sciences. 2010;**54**:1275-1284

[5] Martín-Crespo T, Gómez-Ortiz D, Martín-Velázquez S, Martínez-Pagán P, De Ignacio-San JC, Lillo J, et al. Geoenvironmental characterization of unstable abandoned mine tailings combining geophysical and geochemical methods (Cartagena-La Unión district, Spain). Engineering Geology. 2018;**232**:135-146

[6] Martín-Crespo T, Gómez-Ortiz D, Martínez-Pagán P, Martín-Velázquez S, Esbrí JM, De Ignacio-San JC, et al. Abandoned mine tailings in cultural itineraries: Don Quijote route (Spain). Engineering Geology. 2015;**197**:82-93

[7] Martín-Crespo T, Gómez-Ortiz D, Martínez-Pagán P, De Ignacio-San José C, Martín-Velázquez S, Lillo J, et al. Geoenvironmental characterization of riverbeds affected by mine tailings in the Mazarrón district (Spain). Journal of Geochemical Exploration. 2012;**119-120**:6-16

[8] López M, González I, Romero A. Trace elements contamination of agricultural soils affected by sulphide exploitation (Iberian Pyrite Belt, SW, Spain). Environmental Geology. 2008;**54**:805-818

[9] Cánovas CR, Olías M, Nieto JM, Sarmiento AM, Cerón JC. Hydrogeochemical characteristics of the Odiel and Tinto rivers SW Spain. Factors controlling metal contents. Science of the Total Environment. 2007;**373**:363-382

[10] Aguilar J, Dorronsoro C, Fernández E, Fernández J, García I, Martín F, et al. Soil pollution by a pyrite mine spill in Spain: Evolution in time. Environmental Pollution. 2004;**132**:395-401

[11] López García JA. Estudio mineralógico, textural y geoquímico de las zonas de oxidación de los yacimientos de Fe-Pb y Zn de la Sierra de Cartagena (Murcia) [thesis]. Madrid: Universidad Complutense de Madrid; 1985

[12] JCCM. Ruta de Don Quijote. Un lugar para la aventura [Internet]. 2005. Available from: http://www.castillalamancha.es/sites/default/files/ruta_dqm_mapa_esp.zip [Accessed: 2018-11-30]

[13] Palero FJ, Both RA, Mangas J, Martín-Izard A, Reguilón R. Metalogénesis de los yacimientos de Pb-Zn de la región del Valle de Alcudia (Sierra Morena Oriental). In: García Guinea J, Martínez Frías J, editors. Recursos Minerales de España. Madrid: CSIC; 1992. pp. 1027-1067

[14] Oyarzun R, Lillo J, López-García JA, Esbrí JM, Cubas P, Llanos W, et al. The Mazarrón Pb-(Ag)-Zn mining district (SE Spain) as a source of heavy metal contamination in a semiarid realm: Geochemical data from mine

wastes, soils, and stream sediments. Journal of Geochemical Exploration. 2011;**109**:113-124

[15] Actlabs. Activation Laboratories Ltd. Contract Analytical Services [Internet]. 2018. Available from: http://www.actlabs.com/ [Accessed: 2018-11-30]

[16] Reynolds JM. An Introduction to Applied and Environmental Geophysics. Chichester: Wiley; 1997. 710 p

[17] Sasaki Y. Resolution of resistivity tomography inferred from numerical simulation. Geophysical Prospecting. 1992;**40**:453-464

[18] Gómez-Ortiz D, Martín-Velázquez S, Martín-Crespo T, De Ignacio-San JC, Lillo-Ramos J. Application of electrical resistivity tomography to the environmental characterization of abandoned massive sulphide mine ponds (Iberian Pyrite Belt, SW Spain). Near Surface Geophysics. 2010;**8**:65-74

[19] Loke MH, Acworth I, Dahlin T. A comparison of smooth and blocky inversion methods in 2D electrical imaging surveys. Exploration Geophysics. 2003;**34**:182-187

[20] Cartomur. Infraestructura de datos espaciales de referencia de la región de Murcia [Internet]. 2017. Available from: http://cartomur.imida.es/visorcartoteca/ [Accessed: 2017-02-14]

[21] Montoya-Montes I, Cano-Bermejo I, Sánchez-García MJ, De Ignacio-San JC, Martín-Velázquez S, Gómez-Ortiz D, et al. Formación de cuerpos dunares a partir de lodos mineros: mina de San Quintín (Ciudad Real, España). Geotemas. 2012;**13**:1487-1490

[22] Jiménez Ballesta R, Conde Bueno P, Martín Rubí JA, García Jiménez R. Niveles de fondo geoquímico e influencia del marco geológico en las concentraciones edafogeoquímicas de base de suelos seleccionados de Castilla-La Mancha. Estudios Geológicos. 2010;**66**:123-130

[23] US EPA. Guidelines for water reuse [Internet]. 2004. Available from: https://cfpub.epa.gov/si/si_public_record_report.cfm?Lab=NRMRL&dirEntryId=129543 [Accessed: 2018-11-30]

[24] World Health Organization. Air Quality Guidelines for Europe. Copenhagen: WHO Regional Publications European Series 91; 2000. 288 p

[25] Müller G. Index of geoaccumulation in sediments of the Rhine River. GeoJournal. 1969;**2**:108-118

[26] Taylor SR, McLennan SM. The geochemical evolution of the continental crust. Reviews of Geophysics. 1995;**33**:241-265

[27] Sutherland RA. Bed sediment-associated trace metals in an urban stream, Oahu, Hawaii. Environmental Geology. 2000;**39**:611-626

Chapter 7

Implications of Sediment Geochemistry and Diet Habits in Fish Metal Levels and Human Health Risk

Alice Bosco-Santos and Wanilson Luiz-Silva

Abstract

In this study the concentration of Cd, Cu, Fe, Ni, Pb, and Zn in muscle and liver tissues was compared between four estuarine fish species (*Centropomus parallelus*, *Genidens genidens*, *Diapterus rhombeus*, and *Mugil liza*) to assess contamination levels and the influence of eating habits on metal distribution and human health risk by consumption. In general, liver tissue showed higher metal contents than muscle. Between metals, Fe and Zn contents were relatively higher for both tissues in all analyzed populations. Based on the observations, the variability of metal levels between species is associated with their transfer from the contaminated sediments, where diet habits associated with the substrate result in higher metal accumulation in fish, exerting great influence than bioaccumulation by trophic level. The estimated daily intake (EDI), target hazard quotients (THQ) and the total target hazard quotients (TTHQ), below 1 for all metals on muscle tissues, are suggested the absence of health hazard for the human population. However, high levels of Pb and Zn in liver tissue may endanger predators.

Keywords: contamination, estuaries, fish, metal, risk assessment

1. Introduction

In contaminated environments, risk assessment and exposure to metals by fish consumption are concern issues, because it can become the main route of contaminants to humans. In general, studies focusing on the diagnosis of contaminant levels in fish are focused in the muscle tissue, the main edible part and the major target for metal storage [1–3]. However, little attention, especially in subtropical to tropical environments, has been paid to other tissues, which may provide other information of ecological interest. For instance, analysis of other organs can guarantee the safety of other predators, as bigger fishes and aquatic birds, which feed on the whole specimens.

Tissues as the liver, with highest lipid contents, can warn about recent metal accumulation, since metals can reach it very fast by bloodstream after absorption [4–6]. The literature has considered this organ as responsible for biological detoxification process, where part of the metal might be transferred to less sensitive tissues as the muscle [4, 7].

An estuary is a partially enclosed coastal body of brackish water with one or more rivers or streams flowing into it and with a free connection to the open sea [8]. Estuaries and mangrove ecosystems provide habitats for a large number of organisms and support very high productivity. They are also very densely populated and together with the coast represent about 60% of the world population [9]. It has been increasingly difficult to ignore the consequences of this occupation by industrial and urban activities on aquatic organisms, especially when it comes to disposal of potentially hazardous metals in sediments and water [10]. These dynamic ecosystems have some of the highest biotic diversities and biological production in the world, providing food and shelter to commercially important fish and shellfish species, including shelf species that spend some of their juvenile stages in estuaries [8].

The goal of this study is to characterize the levels of metal in muscle and liver tissues of four estuarine fish species (human-consumable protein) and associate it with different diet habits. Influences of these habits and the physical contaminated compartments on the uptake of metals and their distribution among the tissues are evaluated. The human health risk assessment of metals from fish food intake is also estimated.

2. Study area and methods

2.1 The study area and levels of metal contamination

The Santos-Cubatão Estuarine System comprises a narrow strip of land between the Atlantic Ocean (Santos Bay) and the Serra do Mar mountain chair. The mountain is covered by the Atlantic Forest biome, whereas the estuary hosts a mangrove ecosystem (**Figure 1**). The Morrão River estuary was selected as the area for the present study because it represents the main environment impacted by industrial activities (fertilizer and steel plants and their private harbors) in the area [11, 12]. Sediments in estuaries are generally derived from several sources, which include fluvial, atmospheric, and continental shelf contributions, biological activities, and erosion of the estuarine banks. Sources can vary in the upper and lower reaches of the estuary, with biological inputs generally being more important in higher salinity than in lower salinity region, where terrestrial inputs dominate. In this context, natural sources related to rock weathering, and anthropogenic sources, including industrial and urban activities, may be mixed. The Morrão River estuary is a lower reach from the Santos-Cubatão Estuarine System, where domain sediments come from the weathering and erosion of igneous and metamorphic rocks of the Serra do Mar mountain chain [13]. In addition, the industrial and urban wastes reaching the estuarine system are responsible for the input of trace metals into the environment, which are subsequently incorporated into the sediments [14]. **Table 1** shows the background and contamination metal levels in the study area. The average chemical composition of global shale is shown to comparison. In general, the contamination levels for Cd, Cu, Fe, Pb, and Zn (minor Ni) are an order of magnitude relative to the local geochemical background, which is similar to the average global shale composition (except Cu and Ni, lower) [15].

The Santos-Cubatão Estuarine System has been the object of research since the late 1980s [12, 14, 16–22]. However, data regarding metal concentrations in fish captured from this estuarine system and the related human health risk are scarce. This area hosts the largest industrial park and the first busiest commercial harbor in Latin America and some cities that represent together a population of about 1 million inhabitants. Besides the economic importance for Brazil, this area hosts a

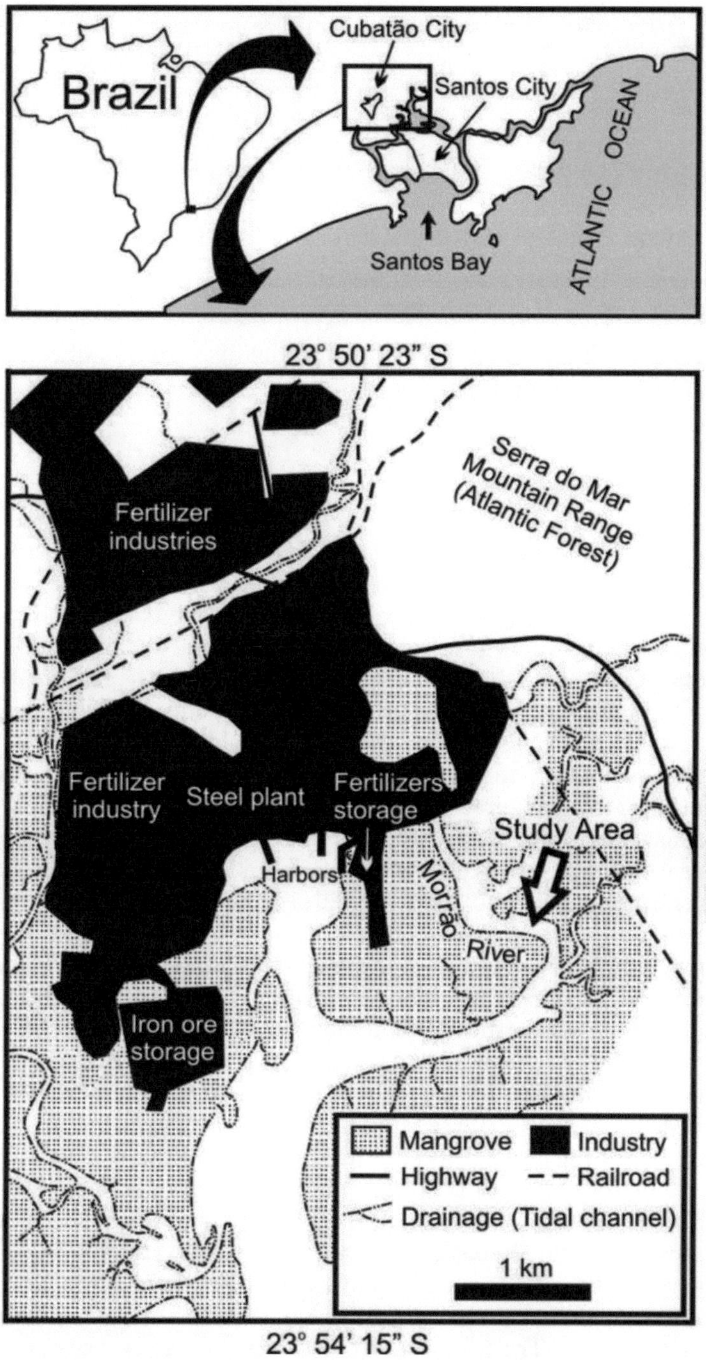

Figure 1.
Study area and sampling points. Source: modified after Gonçalves et al. [11].

preserved mangrove ecosystem and a special part of the Atlantic Forest biome, the latter known as one of the five most important hotspots in the world in terms of endemic species [23]. The chemical study of fish tissues in this highly contaminated area can provide substantial subsidies to improve the understanding about how the health, of physical compartments and dietary habits, affect metal distribution in different species in a subtropical to tropical environment. In addition, the chemical monitoring of a fish community is of great importance since in the area (known as

	Cd	Cu	Fe	Ni	Pb	Zn
Background	0.2 ± 0.1	18.8 ± 0.23	4.69 ± 0.12	27.1 ± 0.60	29.5 ± 3.66	92.3 ± 4.54
Contamination	1.77 ± 0.48	104 ± 27.5	17.6 ± 4.52	50.3 ± 8.73	146 ± 35.9	541 ± 146
Shale	0.3	45	4.72	68	20	95

Table 1.
Concentrations (average ± standard deviation; mg kg^{-1}, except Fe in % weight; n = 4) at 220–260 (representative of the geogenic geochemical background) and 0–20 cm depth (contamination level) based on a bulk sediment core from the Morrão River, Santos-Cubatão estuarine system [12]. Average chemical composition of global shale (mg kg^{-1}) also is shown [15].

"Baixada Santista") thousands of kg of fish and shellfish are caught commercially or sportily for human consumption, besides being a protein source for local aquatic biota including endangered bird species [24, 25].

2.2 Sampling and species habits

A total of 44 fish samples was collected with fishing nets and rods (in the case of the carnivorous species), with the aid of local fishermen, in two campaigns in March 2011, at the Morrão River mangrove banks (**Figure 1**). The fish samples were classified with the aid of Cervigón et al. [26] and Fishbase database [27] in four fish species (*Centropomus parallelus*, *Genidens genidens*, *Diapterus rhombeus*, and *Mugil liza*). These four species together represent 41% of the most fished species of fish in the area. In general, *Genidens genidens* does not have high commercial value, although it is consumed by the local population [24]. The commercial value of *Diapterus rhombeus* is considered medium, and usually specimens longer than 200 mm are marketed [27]. *Mugil liza* is often caught and sold even before sexual maturity is reached, with about 300 mm in length. *Centropomus parallelus* commercial size is between 250 and 300 mm [27].

Centropomus parallelus, *Genidens genidens*, and *Diapterus rhombeus* commonly live in estuaries, and their spatial distribution coincides with the distribution of mangrove ecosystems, which are their main habitat [28–31]. On the other hand, the species *Mugil liza* spawns in the sea, and the young specimens migrate to the estuaries, where they remain until about 8.5 years (*ca.* 500 mm in length) [32]. Therefore, here, young *Mugil liza* specimens only were studied, noting that the body lengths of adult female and male specimens exceed 570 and 500 mm, respectively [33]. The samples of the other three species were mostly adults that had already reached sexual maturity, which occurs when *Centropomus parallelus* is *ca*. 290 mm in total length; *Genidens genidens* female and male specimens are, respectively, *ca.* 133 and 160 mm long and *Diapterus rhombeus* between 80 and 90 mm long [34–37].

According to Rajkowska et al. [38], gender does not exert a significant effect on metal concentrations in most organs of fish, and, therefore, the sex of the specimens was not prioritized here. The most important criterion here was to cover specimens that were near the mangrove regions and resident of the estuary at a certain stage of their life cycle. A summary of the characteristics of the selected species and respective biometric data are shown in **Table 2.**

2.3 Metal analyses

Once collected, the muscle and liver tissues were dissected on a clean surface. The cuts were made with plastic knives to avoid metal contamination [5, 40]. The tissues were freeze-drying, and about 0.25 g (dried in a glass desiccator) of each tissue were digested in 50 mL PFA (Savillex, USA) digestion vessels, by adding

Scientific name (n)[a]	Popular name	Family	Main food items	L (mm)[b]	W (g)[c]
Centropomus parallelus (24)	Fat snook Robalo	Centropomidae	Fish, insects, crustaceans[d]	280 (190–380)	198 (91–980)
Genidens genidens (8)	Catfish Bagre	Ariidae	Algae, benthic crustaceans, mollusks, polychaetes, and bony fish[e]	250 (170–350)	226 (121–800)
Diapterus rhombeus (7)	Mojarra Tainha	Gerreidae	Benthic algae, polychaetes, ostracods, bivalves[f]	250 (170–400)	250 (68–507)
Mugil liza (8)	Mullet Carapeva	Mugilidae	Organic debris and algae[g]	310 (230–470)	382 (154–460)

[a]*n = number of specimens analyzed.*
[b]*L = total length.*
[c]*W = total weight.*
[d]*Tonini et al. [39].*
[e]*Chaves et al. [31].*
[f]*Chaves et al. [29].*
[g]*Fishbase [27].*

Table 2.
Characteristics of the analyzed fish species and specimens.

6 mL concentrated HNO_3 purified by sub-distillation, and then heated for 3 hours at 80 °C (modified after Agah et al.) [41]. The water used was always ultrapure water (18.2 MΩ.cm), obtained from a Milli-Q system (Millipore, USA). All plastic materials were cleaned using a mixed solution of HNO_3 8% and HCl 2% and rinsed with ultrapure water.

The metal concentrations in solutions were obtained by ICP-MS XseriesII (Thermo) equipped with collision cell technology (CCT) in the Geochemistry Laboratory at the University of Campinas. Before the analysis, the instrument was optimized according to the manufacturer recommendations. Lead was measured in normal mode, whereas all other isotopes (^{54}Fe, ^{60}Ni, ^{63}Cu, ^{66}Zn, and ^{114}Cd) were measured using CCT mode. The instrument was calibrated using multi-elemental solutions prepared by mixing the necessary amounts from respective elemental 10 mg L^{-1} (High Purity Standards, USA).

For analytical quality control, we analyzed the certified reference materials DOLT-4 (Dogfish Liver Reference Materials for Trace Metals) and DORM-3 (Dogfish Muscle Certified Reference Material for Trace Metals) produced by the National Research Council of Canada. Calculated recovery was between 83 and 100% (generally better than 90%) considering both certified reference materials, which confirms the good performance of the procedures adopted here (**Table 3**).

2.4 Human health risk assessment

The estimated daily intake (EDI) for Cd, Cu, Fe, Ni, Pb, and Zn (on muscle tissue) was calculated following Eq. (1) (after Saha & Zaman, [42]):

$$\text{EDI} = [(E_F \times E_D \times F_{IR} \times C_F \times C_m)/(W_{AB} \times T_A)] \times 10^{-3} \quad (1)$$

When considering the conservative approach, E_F is the exposure frequency (365 days/year); E_D is the exposure duration (we considered a lifetime of 78 years, average of Brazilians [43]); F_{IR} is the ingestion rate (g/person/day, where we considered 71 g/day, as recorded by the Food and Agricultural Organization [44]); C_F is the conversion factor (0.208) to convert fresh weight to dry weight considering

	Element	Obtained mean ± u	Certified value ± UCRM	Recovery (%)
DOLT-4 (n = 8)	Cd	22.7 ± 0.4	24.3 ± 0.8	94
	Cu	30.3 ± 0.5	31.2 ± 1.1	97
	Fe	1721 ± 38	1833 ± 75	94
	Pb	0.131 ± 0.001	0.16 ± 0.04	84
	Zn	106.5 ± 2.7	116 ± 6	92
DORM-3 (n = 8)	Cd	0.30 ± 0.01	0.29 ± 0.02	100
	Cu	15.2 ± 0.3	15.5 ± 0.63	98
	Fe	327.7 ± 7.1	347 ± 20	94
	Ni	1.06 ± 0.03	1.28 ± 0.24	83
	Zn	46.1 ± 0.7	51.3 ± 3.1	90

Table 3.
Obtained and certified values (mg kg^{-1}) in mass fraction of dry weight in certified reference material DOLT-4 (n = 8) and DORM-3 (n = 8). u = standard deviation/$\sqrt{n}$; UCRM (Uncertainty, certified reference material) = kuc, where uc is the combined standard uncertainty and k is the coverage factor.

approximately 80% moisture content of the fish fillet [45–47]; C_m is the metal concentration in the fish tissue, represented by the mean value of each trace metal at each population analyzed (here an upper confidence limit, UCL95 as show in **Table 4**, was considered as a conservative parameter of population central tendency); W_{AB} is the average body weight for adults (for conservative purpose we assumed the average Brazilian adult body weight for woman, 59.6 Kg [48]); and T_A is the average exposure time for non-carcinogens ($E_F \times E_D$, [46]).

The risk of non-carcinogenic effects was investigated using the target hazard quotient (THQ), which is defined as the ratio between the EDI and the oral reference dose (RfD, mg/kg bw/day) following Eq. (2). This method for estimate THQ considers that for all the potential contaminants, the ingestion dose is equal to the absorbed dose, where cooking has no effect [49]. The RfD represents an estimate of the daily intake oral exposure of the human population that may be continually exposed over a lifetime without an appreciable risk of deleterious effects, and here the USEPA values were applied (0.001 Cd, 0.04 Cu, 0.7 Fe, 0.02 Ni, 0.004 Pb, and 0.3 Zn, [49]).

$$\text{THQ} = \text{EDI}/\text{RfD} \tag{2}$$

As the area of study is contaminated with more than one of the potential contaminants evaluated here, and considering that exposure to two or more pollutants may cause additive effects [47], the cumulative health risk was evaluated by the sum of individuals THQ, expressed as in Eq. (3) as the total THQ (TTHQ).

$$\text{TTHQ} = \text{THQ}(\text{toxicant 1}) + \text{THQ}(\text{toxicant 2}) + \ldots \text{THQ}(\text{toxicant n}) \tag{3}$$

In general, according to the literature [42, 46, 47, 50], values of THQ and TTHQ lower than 1 suggest that adverse hazard of the exposed population to the metals evaluated is not expected.

2.5 Statistical analysis

One way ANOVA followed by parametric correlation (Pearson coefficient) was used to compare metal contents among tissues, species, and the relationships between the trace metal levels in the tissues and the specimens' length and weight. A $p < 0.05$ was considered statistically significant, and all statistical tests were carried out using the OriginPro (v.8.6) graphing and data analysis software. Additionally, principal component analysis (PCA) was used to classify and differentiate samples between different tissues.

3. Results and discussion

3.1 Metal concentration

Mean values, interval of confidence (at 95% level) and concentration ranges of Cd, Cu, Fe, Ni, Pb, and Zn contents, obtained for muscle and liver tissues, from the four fish species with different diet habits are shown in **Table 4**. The content of metals is expressed in dry weight, and wet weight (not shown) should consider approximately 80% moisture content of the fish fillet. Analysis of variance pointed that metal concentrations were significantly different ($p < 0.05$) between tissues for each analyzed species, with higher levels in liver than muscle tissue (except for Pb in *Diapterus rhombeus* and Ni in *Centropomus parallelus*). Iron and Zn were the most abundant elements in both tissues, being at least two orders of magnitude over Cu, Ni, Pb, and Cd contents, particularly in muscle tissue. In liver, Fe concentrations were 8–18 times higher than Zn and Cu, which were between one and four order of magnitude from Ni, Cd, and Pb concentrations.

Higher levels of metals (except Cd) were observed in both tissues (muscle and liver) of noncarnivorous over carnivorous species, alerting for an accumulation pattern related to diet habits. In muscle of *Genidens genidens*, for example, Fe and Zn were 4.3 and 8.7 times higher, respectively, in *C. parallelus* muscle and liver tissues.

Comparing to the global scenario (**Table 5**), Cd, Cu, Ni, and Pb contents in muscle and liver tissues of fish from the Santos-Cubatão Estuarine System were similar to levels found in other impacted environments with similar contaminated sources. However, levels for Fe and Zn stand out being even higher than those reported in the 1980s for Mugilidae, Ariidae, and Centropomidae species [16].

3.2 Metal levels in fish and its relation with contaminated sediment

Two major pathways, or uptake vectors, are responsible for metal incorporation in detritus-feeding aquatic species: (1) ingestion of particles from metal-enriched sediments or (2) uptake by water from particles in suspension [57].

Here, higher levels of metals in tissues from noncarnivorous species suggested that in the impacted scenario of the Santos-Cubatão Estuarine System, the habits associated to the substrate were relevant. The literature suggests that studied area substrate shows a history of contamination related to a steel plant activity since the 1960s, with strong anomalies of Fe and Zn, which reach, in the first 20 cm of the overbank sedimentation, $17.6 \pm 4.6\%$ of Fe and 541 ± 146 mg kg^{-1} of Zn (**Table 1**) [12, 58]. These values are, respectively, *ca.* 5.5 and 8 times higher than those corresponding to the average composition of the shale [15] and are 5.9 (Zn) and 3.7 (Fe) times higher than the pre-industrial values in the study area [14]. As sediments are an item commonly found in omnivorous/detritivorous dietary habits, as *Genidens genidens*, *Diapterus rhombeus*, and *Mugil liza* [29, 31, 39, 59, 60], the data suggest that high Fe and Zn levels observed can be a consequence of the local sediment's intake.

		C. parallelus n = 21 (Carnivorous)			G. genidens n = 8 (Omnivorous)			D. rhombeus n = 8 (Omnivorous)			M. liza n = 7 (Detritivorous)		
		M	L	F	M	L	F	M	L	F	M	L	F
Cd	Mean	0.0013	0.06	**66.8**	0.003	0.24	**24.37**	0.002	0.25	**11.07**	0.001	0.26	**22.34**
	UCL95	0.0016	0.07		0.003	0.33		0.004	0.38		0.0016	0.39	
	Min-Max	0.001–0.004	0.03–0.14		0.001–0.004	0.06–0.44		0.001–0.006	0.05–0.70		0.001–0.02	0.06–0.43	
Cu	Mean	0.61	12.2	**221.3**	1.29	26.1	**35.39**	0.79	10.9	**114.2**	0.79	176	**70.10**
	UCL95	0.63	13.5		1.45	34.0		0.94	12.8		0.98	224	
	Min-Max	0.51–0.73	6.66–21.4		0.92–1.69	6.14–42.1		0.42–1.03	8.20–15.6		0.52–1.15	102–226	
Fe	Mean	7.21	822	**242.3**	39.5	9041	**24.07**	9.71	2468	**14.74**	20.3	2085	**17.87**
	UCL95	8.30	912		44.0	12,517		11.3	3221		25.4	3675	
	Min-Max	0.60–14.6	402–1181		30.9–49.8	2720–18,490		6.96–14.9	312–4257		13.4–30.1	517–3755	
Ni	Mean	0.05	0.06	0.53	0.07	0.13	**5.02**	0.05	0.31	**8.06**	0.36	1.17	**5.02**
	UCL95	0.06	0.07		0.096	0.19		0.07	0.49		0.53	1.79	
	Min-Max	0.05–0.13	0.01–0.11		0.04–0.010	0.05–0.24		0.04–0.50	0.05–0.81		0.03–0.50	0.50–1.79	
Pb	Mean	0.009	0.019	**17.3**	0.07	7.80	**10.62**	0.80	0.34	1.09	0.06	1.13	**83.09**
	UCL95	0.013	0.029		0.09	12.29		1.58	1.04		0.11	1.50	
	Min-Max	0.002–0.03	0.006–0.031		0.02–0.09	0.93–21.8		0.002–2.80	0.002–1.42		0.01–0.22	0.79–1.54	
Zn	Mean	17.2	64.8	**272.8**	80.5	1201	**126.4**	25.2	138	**33.64**	11.8	188	7.85
	UCL95	18.04	69.7		113	1387		32.1	175		13.4	392	
	Min-Max	13.6–20.7	38.3–86.0		29.3–189	934–1658		11.7–39.2	82.3–229		9.3–14.8	60.0–444	

Table 4.
Mean metal contents (mg kg^{-1}), the interval of confidence at 95% level, and the extreme values (minimum and maximum) in mass fraction of dry weight of muscle (M) and liver (L) samples of Centropomus parallelus, Genidens genidens, Diapterus rhombeus, *and* Mugil liza. *Variance values (F) between concentrations in tissues for each element are in bold for $p < 0.05$ indicating significant differences in metal levels. UCL95 was calculated using the USEPA free software ProUCL.*

Family	Region	Tissue	Cd	Cu	Fe	Pb	Ni	Zn	References
Mugilidae	Santos Bay (Brazil)	Muscle	0.06	0.61	—	0.10	—	6.74	[16]
	São Vicente Estuary (Brazil)	Liver	—	7.67 ± 1.22	24.37 ± 55.25	—	4.59 ± 1.38	103.49 ± 18.92	[51]
	Turkey	Muscle	0.66 ± 0.08	4.41 ± 1.67	38.71 ± 18.28	5.32 ± 2.33	—	37.39 ± 6.88	[52]
		Liver	1.64 ± 0.91	202.8 ± 265.8	370.43 ± 252.7	12.59 ± 5.8	—	110.03 ± 34.58	
	Karoun River (Iran)	Muscle	0.84 ± 0.08	—	0.73 ± 0.06	1.75 ± 0.10	—	—	[53]
		Liver	0.52 ± 0.03	—	13.5 ± 0.11	1.03 ± 0.13	—	—	
	Tuzla Lagoon Spring season (Turkey)	Muscle	0.08 ± 0.01	0.62 ± 0.008	11.12 ± 4.13	1.19 ± 0.44	—	60.86 ± 56.10	[6]
		Liver	0.35 ± 0.17	5.74 ± 3.11	305.0 ± 179.60	1.85 ± 0.11	—	28.3 ± 3.71	
Arridae	Paranaguá Estuary (Brazil)	Muscle	—	1.06 ± 0.20	—	—	0.15 ± 0.05	40.42 ± 19.20	[54]
	Santos Bay (Brazil)	Muscle	0.06	0.82		0.09		17.15	[16]
	Gulf of Paria (Venezuela)	Muscle	3.13 ± 1.76	1.35 ± 0.38	—	0.08 ± 0.20	0.12 ± 0.23		[55]
Centropomidae	Vitoria Bay (Brazil)	Muscle	—	0.67 ± 0.07	6.8 ± 0.9	0.03 ± 0.01	0.04 ± 0.04	14.4 ± 0.4	[56]
	Santos Bay (Brazil)	Muscle	0.06	0.28	—	0.09	—	5.68	[16]
Mojarra	Gulf of Paria (Venezuela)	Muscle	4.71 ± 2.31	0.65 ± 0.36	—	0.01 ± 0.02	—	—	[55]

Table 5.
Contents of Cd, Cu, Fe, Pb, Ni, and Zn (mg kg^{-1}, dry weight) in muscle and liver tissues of some specimens of Mugilidae, Ariidae, Centropomidae, and Gerreidae families around the tropical/subtropical world.

Cadmium, Cu, Ni, and Pb anomalies were lower than Fe and Zn in the study area surface sediments (**Table 1**), which can explain the lower Cd, Cu, Ni, and Pb contents observed in the fish tissues (**Table 4**). In fact, other studies have observed that Fe and Zn occur in much higher proportions in bioavailable fractions from the study area sediments than other metals (Fe >>> Zn >> Cu > Pb > Ni >> Cd) [11, 61].

In respect to metal uptake from water, the literature has shown that processes altering redox potential of sediments and chemical forms of metals can promote the flux of metals from sediments to water [62]. Because the diet of noncarnivorous fish consists mainly of benthic organisms (animals and vegetables), and the metal load in the water column of the sampling sites proved to be relatively low [11], the metal levels detected in tissues of these fish seem to reflect the pollution level of the sediment and its biota, rather than the prevailing pollution state of the water.

3.3 Metal in tissues and diet habits

Differences between metal contents in tissues and their correlations with diet habits were depicted by applying principal component analysis (PCA; **Figure 2**). Here only two principal components were considered, and they explained 75.85% (carnivorous species) and 82.91% (detritivorous and omnivorous species) of the total variance in the original data set.

The higher metal intake of noncarnivorous species, especially in respect to Zn and Fe as a consequence of their eating habits and consumed items, is shown in the second principal component in PCA (**Figure 2**). The food items of detritivorous and omnivorous species (**Table 2**) are associated with the contaminated substrate, and thus the small dispersion observed in component 2 (score between −0.5 and 1.0; **Figure 2**) indicates this single source for metals. In contrast, food items of

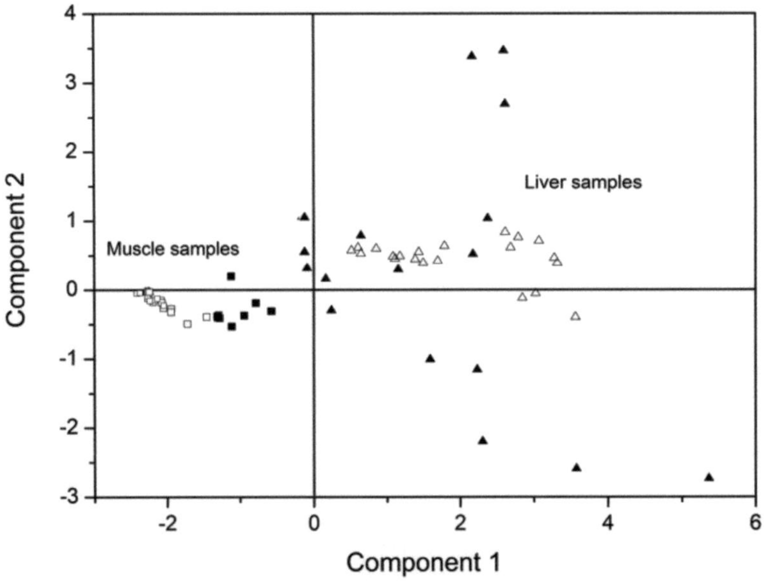

Figure 2.
*Scatter plots of the scores on the first two principal components (explained 75.85% of carnivorous species and 82.91% of detritivorous and omnivorous species) of the total variance in the original data set, obtained using Cd, Cu, Fe, Ni, Pb, and Zn. The black symbols represent the carnivorous species (*C. parallelus*, n = 17), and white symbols represent the omnivorous and detritivorous species (*G. genidens*, *D. rhombeus*, and* M. liza*; n = 20). The squares represent the muscle tissue and the triangles represent the liver tissue.*

carnivorous species (**Table 2**) vary according to their origin. In general, the predator catches its prey in both sedimentary substrate and water column, and these sources show distinct contamination levels [63] that are consequently transferred to the prey differently.

Centropomus parallelus is considered exclusively carnivorous, with 90% of its stomach contents consisting of small animals (fish, crustaceans, polychaetes, and insects), especially other fish that account for 70% of the weight of the content found [39]. This species is classified as sight-feeder, and it tries to capture any moving particle in the water; once in the mouth, it is ingested or rejected, depending on its taste and texture [64]. The variability of prey of possibly different sources is represented by large scores (between −2.7 and 3.6) for the carnivorous species liver samples in component 2 (**Figure 2**). Differences between metal contents in tissues can be observed in **Figure 2**, where two distinct groups represent muscle and liver samples. Component 1 showed, approximately, the same score range for all four species, although tissues from detritivorous and omnivorous species presented muscle and liver domains better separated than carnivorous species. This can be a result of a more varied eating habit of the carnivorous species, and this behavior is better explained by the second principal component (see below).

In general, component 1 showed more disperse liver samples than muscle samples, and it can be a consequence of diet, metabolism, and environmental conditions. Since metals reach the hepatic organ by bloodstream after absorption, the first principal component in respect to liver samples reflects the chemical contrast common in the studied estuarine environment, in which metal levels are heterogeneous spatially [14, 65]. In opposition to the homogeneity expressed by component 1, the metal contents in muscle samples shown in **Figure 2** seem to represent storage of metals during a longer time after internal metabolism and redistribution among tissues. Following uptake into the bloodstream, contaminants are redistributed to high metabolic organs such as the liver, for transformation and detoxification, and then to low metabolic ones, including muscles [4, 66]. Mainly because of the large amount of metallothionein, metal levels in the liver rapidly increase during exposure [6, 7, 67, 68] and are proportional to the levels present in the environment. In contrast, the muscle is the final tissue for metal storage after transformation and excretion of some contaminants [1, 69].

3.4 Health risk assessment

In relation to human consumption, muscle tissues (usually humans do not eat fish liver) require special attention because of toxicity risk. When metal contents exceed the maximum levels established for these contaminants in food, it should be considered unfit for human consumption [5]. In Brazil, the Health Department determined maximum values for fish *in natura* of 0.05 mg kg^{-1} for Cd (with exception for some species as Mojarra and mullet, for which the limit is 0.1 mg kg^{-1}) and 0.3 mg kg^{-1} for Pb [70]. In a less specific legislation, maximum permissible concentrations for Cu (30 mg kg^{-1}), Ni (5 mg kg^{-1}), and Zn (50 mg kg^{-1}) for general food [71, 72] are established. Recalculating the metal contents listed in **Table 4** to wet weight values, most Pb levels in the *Diapterus rhombeus* specimens exceeded the Brazilian legislation limit. Lead contents in the other noncarnivorous species analyzed also deserve attention being extremely close to the limits set forth by the law.

For Cd, Cu, Ni, and Zn, the concentrations are below the limits to offer dangerous to human consumption, and just ingestion of large amounts of fish would

	Cd	Cu	Fe	Ni	Pb	Zn
RfDs	0.001	0.04	0.7	0.02	0.002	0.3
Centropomus parallelus **(number of samples = 21)**						
EDI	5.54×10^{-7}	2.1×10^{-4}	2.8×10^{-3}	2.0×10^{-5}	4.2×10^{-6}	6×10^{-6}
THQ	5.50×10^{-4}	5.33×10^{-3}	4.00×10^{-3}	1.03×10^{-3}	1.06×10^{-3}	2.03×10^{-2}
Genidens genidens **(number of samples = 8)**						
EDI	1.11×10^{-6}	4.88×10^{-4}	1.48×10^{-2}	3.24×10^{-5}	3.75×10^{-5}	3.81×10^{-2}
THQ	1.11×10^{-3}	1.22×10^{-2}	2.12×10^{-2}	1.62×10^{-3}	9.36×10^{-3}	1.27×10^{-1}
Diapterus rhombeus **(number of samples = 7)**						
EDI	1.52×10^{-6}	3.17×10^{-4}	3.82×10^{-3}	2.25×10^{-5}	5.34×10^{-4}	1.08×10^{-2}
THQ	1.52×10^{-3}	7.92×10^{-3}	5.45×10^{-3}	1.12×10^{-3}	1.34×10^{-1}	3.61×10^{-2}
Mugil liza **(number of samples = 8)**						
EDI	5.53×10^{-7}	3.31×10^{-4}	8.56×10^{-3}	1.80×10^{-4}	3.85×10^{-5}	4.50×10^{-3}
THQ	5.53×10^{-4}	8.28×10^{-3}	1.22×10^{-2}	9.01×10^{-3}	9.62×10^{-3}	1.50×10^{-2}

Table 6.
The RfD values and estimated EDI (mg/kg bw/day) and THQ (mg/kg bw/week) by the study area population through the consumption of captured fish from the Santos-Cubatão estuarine system.

present this kind of concern. For Zn, for example, just the consumption of 2.5 kg of *Genidens genidens* muscle tissue would approach the limit. From all the metals evaluated, Cd is of least concern, since the World Health Organization [73] has determined the provisional tolerable weekly Cd intake of 25 μg kg^{-1} body weight, which means an intake of over a thousand pounds of fish to exceed this limit.

The estimated daily intake (EDI) were higher for Zn and Fe, followed by Cu for all analyzed species (**Table 6**), presenting a maximum value of 0.038 mg/kg bw/day for *G. genidens*. The lowest EDI values were reported for Cd followed by Pb and Ni. Using those EDI values and taking in account the USEPA RfDs, the THQ for each metal in each fish population was calculated, and it is shown in **Table 6**. The highest value of THQ was observed for Pb at the *D. rhombeus* population (0.134), while the lowest was observed for Cd at the *C. parallelus* population (0.0005). None of the metals individually exceeded the hazard quotient threshold of 1, implying that the level of daily intake of each examined metal for Brazilian population was lower than that of respective dose. The same happened for TTHO (varying from 0.03 for *C. parallelus* to 0.18 for *D. rhombeus*) that evaluate the cumulative health risks, reinforcing no potential significant health risks. Those observations suggest that the levels of human exposure to the analyzed metals should not cause any deleterious (noncarcinogenic) effect from the intake of fish species from study area (**Table 6**).

3.5 Marine conservation risk assessment

Metal levels in fish liver tissue should be a concern when it comes to marine conservation. In organisms that have aerial respiration (e.g., seabirds and marine mammals), the intake of contaminants is mainly made by food, and biomagnification is usually observed [74]. Although the levels that can cause bad effects to marine organisms vary with the species and metals, the average consumption of 1 kg of *Mugil liza* or *Genidens genidens* studied here, for instance, can expose the predator to high levels of toxic elements, such as Cu and Pb (not essential and without metabolic function), which can exceed 50 and 5 mg kg^{-1}, respectively.

4. Conclusion

This study showed that high Fe and Zn contents found in fish tissues reflect the anomalous concentrations in contaminated sediments. The highest metal contents were found in the liver of noncarnivorous species, protein not normally consumed by humans, but can put at risk predators that eat the whole fish (aquatic birds, fish, and marine mammals). The results showed that fish eating habits, associated with contamination levels in sediments, play an important role in metal uptake. They can exert higher influence on metal levels in fish tissues (muscle and liver) than bioaccumulation by trophic level.

Acknowledgements

We thank the São Paulo State Research Funding Agency—FAPESP (process number 08-11511-8) and the Brazilian National Council for Scientific and Technological Development (process 432922/2016-4)—for financial support. Bosco-Santos thanks the Brazilian Council for Scientific and Technological Development—CNPq—for granting her with a master fellowship.

Author details

Alice Bosco-Santos* and Wanilson Luiz-Silva
Institute of Geosciences, University of Campinas, Campinas, Brazil

*Address all correspondence to: alicebosco@gmail.com

References

[1] Kumar B, Mukherjee DP, Kumar S, Mishra M, Prakash D, Singh SK, et al. Bioaccumulation of heavy metals in muscle tissue of fishes from selected aquaculture ponds in East Kolkata wetlands. Annals of Biological Research. 2011;**2**:125-134

[2] Biswas S, Prabhu RK, Hussain KJ, Selvanayagam M, Satpathy KK. Heavy metals concentration in edible fishes from coastal region of Kalpakkam, southeastern part of India. Environmental Monitoring and Assessment. 2012;**184**:5097-5104

[3] Yildirim Y, Gonulalan Z, Narin I, Soylak M. Evaluation of trace heavy metal levels of some fish species sold at retail in Kayseri, Turkey. Environmental Monitoring and Assessment. 2009;**149**: 223-228

[4] Braunbeck T, Hinton DE, Streit B. Fish Ecotoxicology. 1th ed. Basel, Switzerland: Birkhauser Verlag; 1998

[5] Lima-Júnior RGS, Araújo FG, Maia MF, Pinto ASSB. Evaluation of heavy metals in fish of the Sepetiba and Ilha Grande bays, Rio de Janeiro, Brazil. Environmental Research. 2002;**89**: 171-179

[6] Dural M, Goksu MZL, Ozak AA. Investigation of heavy metal levels in economically important fish species captured from the Tuzla lagoon. Food Chemistry. 2007;**102**:415-421

[7] Oost R, Beyer J, Vermeulen NPE. Fish bioaccumulation and biomarkers in environmental risk assessment: A review. Environmental Toxicology and Pharmacology. 2013;**13**:57-149

[8] Bianchi TS. Biogeochemistry of Estuaries. New York, USA: Oxford University Press; 2007

[9] Alongi DM. Coastal Ecosystem Processes. CRC Marine Science Series, 3. Boca Raton: CRC press; 1998

[10] Birch GF, Olmos MA, Lu XT. Assessment of future anthropogenic change and associated benthic risk in coastal environments using sedimentary metal indicators. Journal of Environmental Management. 2012;**107**: 64-75

[11] Gonçalves WFO, Luiz-Silva W, Machado W, Nizoli EC, Santelli RE. Geochemistry of intertidal sediment pore waters from the industrialized Santos-Cubatão estuarine system, SE Brazil. Anais da Academia Brasileira de Ciências. 2012;**84**:427-441

[12] Luiz-Silva W, Matos RHR, Kristosch GC, Machado W. Variabilidade espacial e sazonal da concentração de elementos-traço em sedimentos do sistema estuarino de Santos-Cubatão (SP). Quimica Nova. 2006;**29**:256-263

[13] Bosco-Santos A, Luiz-Silva W, Dantas EL. Tracing rare earth elements sources in Ucides cordatus crabs by means of 147Sm/144Nd and 143Nd/144Nd isotopic systematics. Journal: Water, Air, & Soil Pollution. 2018;**229**:365. DOI: 10.1007/s11270-018-3990-z

[14] Luiz-Silva W, Machado W, Matos RHR. Multi-elemental contamination and historic record in sediments from the Santos-Cubatão estuarine system-Brazil. Journal of the Brazilian Chemical Society. 2008;**19**: 1490-1500

[15] Turekian KK, Wedepohl KH. Distribution of the elements in some major units of the Earth's crust. Geological Society of America Bulletin. 1961;**72**:175-192

[16] Boldrini CV, Navas-Pereira D. Metais pesados na Baía de Santos e Estuários de Santos e São Vicente: Bioacumulação. Ambiente. 1987;**1**: 118-127 (In Portuguese)

[17] Gutberlet J. Cubatão: Desenvolvimento, Exclusão Social e Degradação Ambiental. 1st ed. São Paulo, Brazil: Editora da Universidade de São Paulo; 1996

[18] Hortellani MA, Sarkis JES, Bonetti J, Bonetti C. Evaluation of mercury contamination in sediments from Santos-São Vicente estuarine system, São Paulo state, Brazil. Journal of the Brazilian Chemical Society. 2005;**16**: 1140-1149

[19] Hortellani MA, Sarkis JES, Abessa DMS, Sousa ECPM. Avaliação da contaminação por elementos metálicos dos sedimentos do estuário Santos-São Vicente. Quimica Nova. 2008;**31**:10-19

[20] Bordon ICAC, Sarkis JES, Gobatto GM, Hortellani MA, Peixoto CM. Metal concentration in sediments from the Santos estuarine syste A recent assessment. Journal of the Brazilian Chemical Society. 2011;**22**: 1858-1865

[21] Azevedo JS, Braga ES, Favaro DT, Perretti AR, Rezende CE, Souza CMM. Total mercury in sediments and in Brazilian Arridae catfish form two estuaries under different anthropogenic influence. Marine Pollution Bulletin. 2011;**62**:2724-2731

[22] Luiz-Silva W, Matos RHR, Kristosch GC. Geoquímica e índice de geoacumulação de mercúrio em sedimentos de superfície do estuário de Santos-Cubatão (SP). Quimica Nova. 2002;**25**:753

[23] Myers N, Mittermeier RA, Mittermeier CG, Fonseca GAB, Kent J. Biodiversity hotspots for conservation priorities. Nature. 2000;**403**:853-858

[24] Cetesb report. Available from: http://www.cetesb.sp.gov.br/Agua/relatorios/rel_sist_estuarino/relatorio.zip [Accessed: January 05, 2013]

[25] Olmos F, Silva RS. Breeding biology and nest site characteristics of the scarlet Ibis in southeastern Brazil. Waterbirds. 2001;**24**:58-67

[26] Cervigón F, Cipriani R, Fischer W, Garibaldi L, Hendrickx M, Lemus AJ, et al. Guia de campo de las espécies comerciales marinas y de águas salobres de la costa septentrional de Sur America. Roma: Organizacion de las Naciones Unidas para la Agricultura y la Alimentacion; 1992

[27] Fishbase. Available from: http://www.fishbase.org [Accessed January 04, 2011]

[28] Gilmore RG, Donohoe J, Cooke DW. Observations on the distribution and biology of east-Central Florida populations of the common Snook, *Centropomus undecimalis*. Florida Scientist. 1983;**46**:313-336

[29] Chaves PTC, Otto G. Aspectos biológicos de *Diapterus rhombeus* (Cuvier) (Teleostei, Gerreidae) na Baía de Guaratuba, Paraná, Brasil. Revista Brasileira de Zoologia. 1998;**15**:289-295

[30] Schmidt TCS, Martins IA, Reigada ALD, Dias JF. Taxocenose de bagres marinhos (Siluriformes, Ariidae) da região estuarina de São Vicente, SP, Brasil. Biota Neotropica. 2008;**8**:73-81

[31] Chaves PTC, Vendel AL. Aspectos da Alimentação de Genidens Genidens (Valenciennes) (Siluriformes, Ariidae) na Baía de Guanabara, Paraná. Revista Brasileira de Zoologia. 1996;**13**:669-675

[32] Garbin T, Castello JP, Kinas PG. Age, growth, and mortality of the mullet Mugil liza in Brazil's southern and southeastern coastal regions. Fisheries Research. 2014;**149**:61-68

[33] Albieri RJ, Araújo FG. Reproductive biology of the mullet Mugil liza (Teleostei: Mugilidae) in a tropical Brazilian bay. Zoologia. 2010;**27**:331-340

[34] Taylor RG, Whittington JA, Grier HJ, Crabtree RE. ge, growth, maturation, and protandric sex reversal in common Snook, Centropomus undecimalis, from the east and west coasts of South Florida. Fisheries Bulletin. 2000;**98**:612-624

[35] Mazzoni M, Tanji SB. Reproductive biology of Genidens genidens, a catfish from the Maricá lagoon, RJ. Ciência e Cultura. 2000;**52**:121-126

[36] Barbieri LR, Santos RP, Andreara JV. Reproductive biology of the marine catfish, Genidens genidens in the Jacarepaguá Lagoon system. Environmental Biology of Fishes. Rio de Janeiro, Brazil; 1992;**35**:23-35

[37] Costa MR, Albieri RJ, Neves LM, Santos ABL, Araujo FG. Distribution and size of the mojarra *Diapterus rhombeus* (Cuvier) (Actinopterygii, Gerreidae) in a southeastern Brazilian bay. Brazilian Journal of Oceanography. 2012;**60**:199-207

[38] Rajkowska M, Protasowicki M. Distribution of metals (Fe, Mn, Zn, Cu) in fish tissues in two lakes of different trophy in northwestern Poland. Environmental Monitoring and Assessment. 2013;**185**:3493-3502

[39] Tonini WCT, Braga LGT, Vila Nova DLD. Nova dieta de juvenis do robalo Centropomus parallellus Poey, 1860 no sul da Bahia, Brasil. Boletim do Instituto de Pesca. 2007;**33**:85-91

[40] Karadede H, Oymak SA, Unlu E. Heavy metals in mullet, Liza abu, and catfish, *Silurus triostegus*, from the Atatürk dam Lake (Euphrates), Turkey. Environment International. 2004;**30**: 183-188

[41] Agah H, Leermakers M, Elskens M, Fatemi SMR, Baeyens W. Accumulation of trace metals in the muscle and liver tissues of five fish species from the Persian Gulf. Environmental Monitoring and Assessment. 2009;**157**: 499-514

[42] Saha N, Zaman M. Evaluation of possible health risks of heavy metals by consumption of foodstuffs available in the central market of Rajshahi City, Bangladesh. Environmental Monitoring and Assessment. 2013;**185**:3867-3878

[43] Instituto Brasileiro de Geografia e Estatística. Available from: https://www.ibge.gov.br/apps/populacao/projecao/ [Accessed: July 2019]

[44] Food and Agricultural Organization (FAO), 2008. Statistics Division. Food Security Statistics. Food Consumption. Available from: http://www.fao.org/es/ESS/faostat/foodsecurity/index_en.htm [Accessed Januray 05, 2018]

[45] Rahman MS, Molla AH, Saha N, Rahman A. Study on heavy metals levels and its risk assessment in some edible fishes from Bangshi River, Savar, Dhaka. Bangladesh. Food Chemistry. 2012;**134**(4):1847-1854

[46] Saha N, Mollah M, Alam M, Rahman MS. Seasonal investigation of heavy metals in marine fishes captured from the bay of Bengal and the implications for human health risk assessment. Food Control. 2016;**70**: 110-118

[47] Mohammad M, Akbari E, Amrane A, Jafari SJ, Eslami H, Hoseinzaded E, et al. Health risk assessment of heavy metal intake due to fish consumption in the Sistan region, Iran. Environmental Monitoring and Assessment. 2017;**189**:583

[48] IBGE. Censo demográfico 2010. Características da população e dos

domicílios: resultados do universo. Rio de Janeiro: IBGE. p. 2011

[49] USEPA, IRIS (Integrated Risk Information System) Compares IRIS Values. Available from: http://cfpub.epa.gov/ncea/iris/compare.cfm [Accessed June 03, 2013]

[50] USEPA, Risk Assessment-Regional Screening Levels (RSLs)-User's Guide. Available from: https://www.epa.gov/risk/regional-screening-levels-rsls-users-guide#target [Accessed June 09, 2019]

[51] Carmo CA, Abessa DMS, Neto JGM. Metais em águas, sedimentos e peixes coletados no estuário de São Vicente-SP, Brasil. O Mundo da Saúde. 2011;**35**:64-70

[52] Canli M, Atli G. The relationships between heavy metal (Cd, Cr, Cu, Fe, Pb, Zn) levels and the size of six Mediterranean fish species. Environmental Pollution. 2003;**121**:129-136

[53] Sary AA, Beheshti M. Cadmium, iron, lead and mercury bioaccumulation in abu mullet, liza abu, different tissues from Karoun and Karkheh Rivers, Khozestan, Iran. Bulletin of Environmental Contamination and Toxicology. 2012;**88**:158-161

[54] Angeli JLF, Trevizani TH, Ribeiro A, Machado EC, Figueira RCL, Markert B, et al. Arsenic and other trace elements in two catfish species from Paranaguá estuarine complex, Paraná, Brazil. Environmental Monitoring and Assessment. 2013;**185**:8333-8342

[55] Lemus M, Castañeda J, Chung K. Heavy metals in fish and invertebrates from the Gulf of Paria, Venezuela. Revista Internacional de Contaminacion Ambiental. 2014;**30**:137-142

[56] Souza IC, Duarte ID, Pimentel NQ, Rocha LD, Morozesk M, Bonomo MM, et al. Matching metal pollution with bioavailability, bioaccumulation and biomarkers response in fish (*Centropomus parallelus*) resident in neotropical estuaries. Environmental Pollution. 2013;**180**:136-144

[57] Luoma SN. Can we determine the biological availability of sediment-bound trace elements? Hydrobiologia. 1989;**176**(177):379-396

[58] Bosco-Santos A, Luiz-Silva W, Machado W, Sanders C, Dantas EL, Silva-Filho EV, et al. Fractionation of rare earth and other trace elements in crabs, *Ucides cordatus*, from a subtropical mangrove affected by fertilizer industry. Journal of Environmental Sciences. 2016;**54**:69-76

[59] Seixas TG, Kehrig HA, Moreira I, Malm O. Selênio em tecidos de quatro organismos marinhos da Baía de Guanabara-RJ. Tropical Oceanography. 2005;**33**:207-222

[60] Elliff CI, Tutui S, Ldos S, Souza MR, Tomás ARG. Biologia populacional da carapeba *Diapterus rhombeus* (CUVIER, 1829) (Teleostei: Gerreidae) no estuário de Santos (SP). Boletim do Instituto de Pesca. 2013;**38**:411

[61] Nizoli EC, Luiz-Silva W. Seasonal AVS–SEM relationship in sediments and potential bioavailability of metals in industrialized estuary, southeastern Brazil. Environmental Geochemistry and Health. 2012;**34**:263-272

[62] Fan W, Wang W, Chen J. Geochemistry of Cd, Cr, and Zn in highly contaminated sediments and its influences on assimilation by marine bivalves. Environmental Science & Technology. 2002;**36**:5164-5171

[63] Bervoets L, Blust R. Metal concentrations in water, sediment and

gudgeon (*Gobio gobio*) from a pollution gradient: Relationship with fish condition factor. Environmental Pollution. 2003;**126**:9-19

[64] Bórquez A, Cerqueira VR. Feeding behavior in juvenile Snook, Centropomus undecimalis. I–individual effect of some chemical substances. Aquaculture. 1998;**169**:25-35

[65] Sanders LM, Luiz-Silva W, Machado W, Sanders CJ, Marrota H, Enrich-Prast A, et al. Rare earth element and radionuclide distribution in surface sediments along an estuarine system affected by fertilizer industry contamination. Water, Air, and Soil Pollution. 2013;**224**:1742-1747

[66] Hauser-Davis RA, Bastos FF, Oliveira TF, Ziolli RL, Campos RC. Fish bile as a biomarker for metal exposure. Marine Pollution Bulletin. 2012;**64**: 1589-1595

[67] Kasper D, Palermo EFA, Dias ACMI, Ferreira GL, Leitão RP, Branco CWC, et al. Mercury distribution in different tissues and trophic levels of fish from a tropical reservoir, Brazil. Neotropical Ichthyology. 2009;**7**:751-758

[68] Jezierska B, Witeska M. In: Twardowska I, Allen HE, Haggblom MM, Stefaniak S, editors. Soil and Water Pollution Monitoring Protection and Remediation. Berlin: Springer; 2006

[69] Catsiki VA, Strogyloudi E. Survey of metal levels in common fish species from Greek waters. Science of the Total Environment. 1999;**237–238**:387-400

[70] Agência Nacional de Vigilância Sanitária do Ministério da Saúde, Brasil. Resolução-RDC no42, Diário Oficial da União, 168, 2013

[71] Ministério da Saúde, Brasil. Decreto n° 55.871, 26 de março de 1965

[72] Agência Nacional de Vigilância Sanitária do Ministério da Saúde, Brasil. Portaria n° 685, 27 de agosto de 1998

[73] World Health Organization. Evaluation of certain food additives and contaminants, Technical Report Series no 960, 2011

[74] Gray JS. Biomagnification in marine systems: The perspective of an ecologist. Marine Pollution Bulletin. 2002;**45**: 46-52